A Mathematician's Survival Guide

Graduate School and
Early Career Development

A Mathematician's Survival Guide

Graduate School and
Early Career Development

Steven G. Krantz

American Mathematical Society

2000 *Mathematics Subject Classification.* Primary 00–01.

For additional information and updates on this book, visit
www.ams.org/bookpages/gscm

Library of Congress Cataloging-in-Publication Data
Krantz, Steven G. (Steven George), 1951–
  A mathematician's survival guide : graduate school and early career development / Steven G. Krantz.
     p. cm.
  Includes bibliographical references and index.
  ISBN 0-8218-3455-X (alk. paper)
  1. Mathematics—Vocational guidance. 2. Mathematics—Study and teaching (Higher) I. Title.

QA10.5.K73 2003
510′.23—dc21   2003051889

Copying and reprinting. Individual readers of this publication, and nonprofit libraries acting for them, are permitted to make fair use of the material, such as to copy a chapter for use in teaching or research. Permission is granted to quote brief passages from this publication in reviews, provided the customary acknowledgment of the source is given.
  Republication, systematic copying, or multiple reproduction of any material in this publication is permitted only under license from the American Mathematical Society. Requests for such permission should be addressed to the Acquisitions Department, American Mathematical Society, 201 Charles Street, Providence, Rhode Island 02904-2294, USA. Requests can also be made by e-mail to reprint-permission@ams.org.

© 2003 by the American Mathematical Society. All rights reserved.
The American Mathematical Society retains all rights
except those granted to the United States Government.
Printed in the United States of America.
∞ The paper used in this book is acid-free and falls within the guidelines
established to ensure permanence and durability.
Visit the AMS home page at http://www.ams.org/
10 9 8 7 6 5 4 3 2 1   08 07 06 05 04 03

To the many fine mentors that I had in graduate school. They set me on the right path, and guided me to the finish. They launched me on a wonderful career, and gave me strength when I needed it. I owe them a great deal.

Contents

Preface	xi

Part 1. Getting Ready for Graduate School

Chapter 1.	Heading Off to Graduate School	3
§1.1.	Impressions of Life after College	3
§1.2.	What to Look For in the Future	7
Chapter 2.	Preliminaries	9
§2.1.	How to Prepare for Graduate School	9
§2.2.	The Undergraduate Research Experience	13
§2.3.	Summary of the Optimal Qualifications for Graduate School	15
§2.4.	What About Those GRE's?	16
§2.5.	How to Choose and Apply to a Graduate School	18
§2.6.	Special Considerations for Underrepresented Groups	25
§2.7.	What About My English?	26
§2.8.	How to Pay for Graduate School	27

Part 2. Essential Elements of a Graduate Education

Chapter 3.	Pre-Thesis Work	33
§3.1.	Your Early Courses and Preparation for the Quals	33
§3.2.	How Am I Expected to Perform in My Graduate Classes?	37
§3.3.	The Qualifying Exams	38

§3.4.	When Should I Take the Quals?	40
§3.5.	How Am I Expected to Perform on the Qualifying Exams?	41
§3.6.	What Will My Fellow Graduate Students Be Like?	42
§3.7.	Am I Supposed to Work All of the Time?	43
§3.8.	Should I Attend the Colloquium?	44
§3.9.	The Foreign Language Requirement	45
§3.10.	Should I Join a Professional Society?	47
§3.11.	My Duties as a Teaching Assistant	48
§3.12.	How Will I Be Trained for My TA Duties?	49
§3.13.	Organized Labor Among Graduate Students	50
§3.14.	The Research Assistantship	51
§3.15.	The Departmental Staff	51

Chapter 4. Thesis Work — 53

§4.1.	How Do I Choose a Thesis Advisor?	53
§4.2.	Who Is Eligible to Be My Thesis Advisor?	58
§4.3.	How Do I Find a Thesis Problem?	59
§4.4.	My Relationship with My Thesis Advisor	61
§4.5.	How Do I Work on My Thesis Problem?	63
§4.6.	How Do I Write Up My Thesis?	66
§4.7.	My Thesis Committee and My Defense	70

Part 3. Sticky Wickets

Chapter 5. Practical Difficulties — 75

§5.1.	Should I Hold an Extra Job While I Am a Graduate Student?	75
§5.2.	What If I Can't Solve My Thesis Problem?	76
§5.3.	How Much is Enough for a Thesis?	77
§5.4.	Why Does a Graduate Student Leave the Program?	78
§5.5.	I Don't Seem to Know Anything!	79
§5.6.	Why Does Everyone Else Appear to Be Succeeding?	80

Chapter 6. Moral Difficulties — 83

§6.1.	Am I in Competition With My Fellow Graduate Students?	83
§6.2.	What Does it Mean to Be "Fired" by My Thesis Advisor?	85
§6.3.	Academic Integrity	86
§6.4.	Intimacy with Members of the Mathematics Department	88

Part 4. Post-Graduate-School Existence

Chapter 7.	Life After the Thesis	93
§7.1.	Should I Publish My Thesis?	93
§7.2.	What is a Tenure-Track Job?	95
§7.3.	Looking for a Job	98
§7.4.	Nonacademic Jobs	105
§7.5.	The Job Interview	109
§7.6.	The Life of An Assistant Professor	111
§7.7.	The Tenure Clock	117
§7.8.	What Will Be My Teaching Load?	117
§7.9.	When Do I Get a Sabbatical?	119
§7.10.	What Kind of Money Can I Make as a Professor?	121
Chapter 8.	Afterthoughts	125
§8.1.	Research vs. Teaching	125
§8.2.	How Do I Keep My Research Program Alive?	126
§8.3.	Collaborators	127
§8.4.	Publish or Perish	127
§8.5.	Do All Assistant Professors Become Associate Professors? Do All Associate Professors Become Full Professors?	128
§8.6.	What If I Don't Get Tenure?	130
§8.7.	What If Graduate School Was a Big Mistake?	131
§8.8.	What If I Only Want a Master's Degree?	132
§8.9.	Rounding Out the Graduate School Experience	133

Part 5. The Elements of Mathematics

Chapter 9.	The Mathematics I Need to Know	137
§9.1.	Real Analysis	138
§9.2.	Complex Analysis	144
§9.3.	Geometry/Topology	149
§9.4.	Algebra	158
§9.5.	How Do All of These Subjects Fit Together?	165
Glossary		169
APPENDIX I: The Administrative Structure of a Mathematics Department and a University		191

APPENDIX II: The Academic Ranks 197

APPENDIX III: The Academic Composition of a Mathematics
    Department 201

APPENDIX IV: A Checklist for Graduate School 205

Bibliography 211

Index 215

Preface

When I was a first-year graduate student, I met once per week with a friend to talk about algebraic topology. We were reading Marvin J. Greenberg's fine book [GRE], trying to teach ourselves the subject as we worked through it. Of course algebraic topology is a lovely blend of algebra, topology, logic, category theory, and geometry. It makes considerable demands on the neophyte. And, even though we were both quite able students with excellent backgrounds, we frequently found ourselves staring at each other in frustration and saying, "Just what have we been doing for the past four years? Why don't we seem to know anything?"

In spite of the fact that we had both been the top math majors in our undergraduate programs, my friend and I frequently felt lost. As though we were grossly ill-prepared, and had no sense of what we were doing.

This is a common lament. It is the nature of the American undergraduate education in mathematics to be disjointed. The vast panorama of mathematics is divided up into subject areas and segments, and there is really no venue for the student to become aware of all the beautiful cross-fertilization that is both possible and necessary in the actual *use* of mathematics. Certainly any good research mathematician finds himself/herself grabbing results from algebra, analysis, Lie group theory, logic, geometry, differential equations, and so on without thinking of these as separate resources. Mathematics is one great tapestry, and those who truly understand this fundamental fact are the ones who are on the cutting edge and who can create new mathematics.

More generally, there is no obvious way for the bright undergraduate to find out what graduate school is all about, what the choices are, what the path through graduate school is, and what the steps and pitfalls are along the way.[1] In many instances the journey through graduate school is one of the blind being led by nobody.

It is in the nature of what it means to be a first-year graduate student to struggle with identifying the immediate goals and to determine the means of achieving those goals. Some graduate programs are better than others at helping beginning graduate students to effect the transition from a tyro who has taken a variety of math classes into a working mathematician. The process of becoming a mathematician is a *synthesis*, both of ideas and of technique. It is a process by which one makes the transition from a clever prodigy to a working scholar. This passage is a struggle; indeed, it *must* be a struggle, for we all know (thanks to Euclid) that there is no royal road to learning.

The purpose of the present book is to aid the budding graduate student in understanding the graduate education process and the world beyond it, as well as to aid in understanding the material that needs to be learned. The book has nine vectors in it:

- **(i)** to acquaint the student with the steps necessary to become a mathematician who has the confidence and the resources to actually *use* mathematics in a creative way;
- **(ii)** to familiarize the student with the different steps and features of a graduate program;
- **(iii)** to identify some of the pitfalls in graduate study;
- **(iv)** to describe the basic body of material that a graduate student must learn in order to pass the qualifying exams and move on to writing a thesis;
- **(v)** to show how to harness the tools necessary to write a good thesis;
- **(vi)** to explain how to find a thesis advisor and a thesis problem;
- **(vii)** to describe the process of writing the thesis;
- **(viii)** to explain how one gets an academic job and becomes established in the profession,

  and

- **(ix)** to discuss the study techniques and mental disciplines necessary to effect **(i)**–**(viii)**.

---

[1]Some precocious undergraduates will hang out with the graduate students, and will glean thereby a few ideas about graduate school—just by osmosis.

At the very end of the book, I will provide a very general description of the parts of mathematics that you will need to know for the qualifying exams. I will walk the reader through every step of a typical American graduate program, and then I will describe what happens to you after you leave school (and what you need to know to succeed). In other words, I will treat the basic ingredients of a solid graduate mathematics education and of the first few years in the mathematics profession.

Most mathematics graduate programs in this country conform to a familiar paradigm. The ingredients have been largely standardized. First, virtually every program has a set of so-called "qualifying exams".[2] These are tests to check that the prospective Ph.D. student has mastered certain basic material (covered elsewhere in this book). The topics may include measure theory, basic functional analysis, abstract algebra, topology, differential geometry, and complex analysis (the particulars will vary from school to school, but this list captures the spirit of what is often required).

Along the way, the student will take courses. There are courses to prepare for the quals, courses to bridge the gap (after the quals) to thesis work, and (seminar) courses at the research level.

After the quals comes a period in which the student seeks a thesis advisor and a thesis topic. At some institutions this process is aided by a collection of reading courses in which the student becomes closely acquainted with a selection of faculty and their research.[3] In some programs the student will be eased into participation in some research seminars and will thereby become acquainted with some of the faculty and their work. At other schools (such as my own) the student is required to present a "Minor Oral Exam" and a "Major Oral Exam"; these give the student some experience reading the literature, giving an oral presentation, and getting into the research groove. The faculty advisor to the Major Oral often becomes the thesis advisor. More will be said about these processes later in the book.

Finally, the student must write and defend a thesis. Unfortunately, this last step is not always a salubrious experience for all concerned. The work preceding the thesis is difficult and demanding, but it is still in the nature of rote class work. The thesis demands originality, tenacity, courage, and luck. And a good thesis advisor to help one over the rough spots. Unfortunately, this process is complicated and unpredictable, and more than a few promising candidates fall through the cracks. It is hoped that this book

---

[2] Qualifying exams are also known as "quals", "prelims", or "generals".

[3] In fact, this process is loosely analogous to the device used in medical schools and lab sciences known as "rotations". In that system, students spend a few months in each of several laboratories in order to acquaint themselves with a number of professors, their labs, and their styles. They subsequently pick a lab and a professor with which to do the thesis work.

will provide some guidance to help you make it through this possibly trying period of your education.

An ancillary, but extremely important, portion of the graduate education process is teaching experience. Most graduate students serve as Teaching Assistants (TAs). Typically, the TA will teach some recitation or problem sessions. In some cases the TA will have complete charge of certain classes—writing the syllabus, writing the exams, and assigning the grades. These days such experience is of vital importance. The graduate student seeking a first job and having no teaching experience is simply out of luck. Everyone must have a teaching dossier (more on this below). We will certainly spend some time discussing the role of teaching in the life of the graduate student.

This book will treat every aspect of the life of a graduate student: how to prepare for graduate study; how to select a graduate school; how to get into graduate school; what is expected of you in graduate school; the steps in a graduate education, the qualifying exams, writing a thesis, and so forth. And then, after you have done all that, how do you get a job? What will be the nature of that job, and what will be expected of you? We shall leave no stone unturned in telling you how to become a mathematician.

The book contains an extensive Glossary, both to aid in reading the book and in dealing with academic patois. There are URLs (i.e., web addresses) for many of the terms in this Glossary.

I am an academic mathematician, and that is the life and career path that I know best. I confess that most of this book is directed to people who want to travel that same road, but I have also included material for those who wish to terminate their education with a Master's degree. I discuss at some length the job opportunities in industry and with the government. And I speak about jobs at junior colleges, comprehensive universities, Liberal Arts colleges, teaching colleges, and nonresearch institutions. Most of my remarks are directed toward students being educated in America, but, again, there is some material for students coming from other countries.

It is my fervent belief, well-supported by experience, that the main reason that people often fail at tasks or programs that they set for themselves is that they never figure out what it was that they were supposed to be doing. The Ph.D. program in mathematics is a multistep, fairly complex process. There are many junctures at which one could lose track and not get the right mentoring or advice. The purpose of this book is to provide some objective reference material, presented in an accessible but authoritative tone, to aid in the graduate education process.

It would be natural for any reader to ask why this author feels qualified to write a book such as this one. After all, I am neither a Fields Medalist nor a member of the National Academy of Sciences. Who am I to tell other people how to live? Well, I have been a mathematician for thirty years. I have worked in many different types of mathematics departments, served on every possible kind of committee, and now Chair a math department. I have mentored quite a few graduate (Ph.D. and Master's) students. I regularly teach our graduate course on how to teach (using the textbook *How to Teach Mathematics* by this author)—required of all our graduate students. I make it my business to monitor the progress of *all* of our graduate students and to make sure they get through the program. I *must* be concerned with the issues being treated here. And I simply want to share the benefits of my experience. A permanent member of the Institute for Advanced Study, whatever that person's mathematical virtues may be, will *not* have had many of these life experiences and therefore would perhaps have less to say about these matters. I believe that I can make a useful contribution.

It is a pleasure to thank Jerry Alexanderson, Lynn Apfel, Steven Bell, Harold Boas, Robert Burckel, Joe Cima, John B. Conway, John D'Angelo, Fausto Di Biase, Craig Evans, Jerry Folland, Ron Freiwald, Loukas Grafakos, Anne-Katrin Herbig, Seth Howell, Lauren Kennell, Leonid Kovalev, Lisa Kuehne, Mohan Kumar, John McCarthy, Jeff McNeal, David Opela, Harold R. Parks, Richard Rochberg, Suzanne Tourville, Edward Wilson, and Hung-Hsi Wu for reading various drafts of this manuscript and offering useful comments and suggestions. Don Sarason provided detailed information about the Berkeley graduate program. David Bressoud and Kimberly Pearson taught me about Liberal Arts colleges and comprehensive universities. Edward Dunne and John Ewing of the AMS gave the manuscript a particularly careful read, and contributed many ideas, criticisms, and important bits of information. Gil Poulin was a terrific production editor for the AMS; his sharp eye and sense of language helped to sharpen my prose at many junctures. The idea for this book grew out of conversations with Don Albers of the MAA, and I thank him for the inspiration. All remaining errors and foolishness are of course the sole responsibility of the author.

I look forward to comments and criticisms from the readership of this book. I want to continue to improve it so that it will be a useful tool for generations to come.

Part 1

Getting Ready for Graduate School

Chapter 1

Heading Off to Graduate School

1.1. Impressions of Life after College

OK, it is your first day at your new graduate school. You have moved into your dorm room or other residential digs. You clutch in your hand a letter of admission, together with (most likely) a promise of financial support. The letter contains some vague statements about teaching duties, about "making normal progress", and about the number of years in the program (as well as about the number of years the financial support will last). Now *what are you supposed to do?*

When I was a beginning graduate student there was a mandatory first meeting at which a secretary took a group photograph and some professors told us a few things about the program. The main thing they had to say was that the money would run out in three years so we had better get to work.

These days, most graduate programs will have an elaborate and well-developed orientation procedure—in fact one that may involve several sessions and stretch over a few days. You may be expected to assume Teaching Assistant (abbreviated TA) duties right away (see Sections 3.11, 3.12), and there will probably be some activities to acquaint you with the duties of a TA, and what will be expected of you in that regard. There will likely be some handouts telling you when qualifying exams are given and how you sign up for them. If you are lucky, your program will assign you a pre-thesis advisor to look after you and help you choose your classes (and help you handle any unanticipated difficulties) until you find a subject area and choose a thesis advisor.

Everyone knows that four years is the standard period for an undergraduate education in this country.[1] For a Ph.D. program there is no standard length of time to completion. Even though my graduate program was pegged at three years, most are pegged at five. This means that they would like you to finish (and the money will probably run out) in five years. But different people are different, and every student's background and preparation is a bit different. You may require six years, and that is probably OK. Some students even require seven, and that is probably OK. Please understand that your graduate program and university will be investing a lot in you, and they want you to finish. They are ready and willing to support you as much as the limits of their program will allow.[2]

As your program develops, keep in touch with the Graduate Director and with your thesis advisor to make sure you are making good progress. The rest should take care of itself.[3]

If you are in graduate school, then you are probably at least 22 years old. You are now an adult and you should look after yourself. Supposing that this is your first day in graduate school, I admonish you to take charge of the situation. Figure out where the math building is, go there, and introduce yourself to the Chairperson's secretary and the Graduate secretary. If the Graduate Chairperson[4] is around, shake hands and introduce yourself. Ask

---

[1] Five hundred years ago in England the undergraduate education was set at three years just because the gentry felt that that was a nice length of time for junior to be away from home. The tradition lives on to this day at Oxford and Cambridge.

[2] In some disciplines—literature is one of them—it is nearly impossible for the student to finish the Ph.D. program in any pre-specified period of time. It is quite common for students to leave school after five years or so—the support money will have run out anyway—and go off to a job, raising a family, and all the rest. The student, if s/he is tenacious and lucky, will work on the thesis during evenings and weekends and ultimately finish—sometimes after ten or fifteen years or more.

[3] When I showed up in graduate school—in the early 1970's—it was the end of a remarkable period in American graduate education. For in the late 1950's and the 1960's, in the wake of the "Sputnik era", it was quite common for a student in the sciences and engineering to go to graduate school, pass the quals, be assigned a thesis problem, and then go off to a job—after just a couple of years of graduate work! There was *that great a rush* to produce Ph.D.'s and to get people off into the work force. The expectation was that, with the aid of a few trips back to graduate school to consult with the thesis advisor, the student would complete the Ph.D. thesis during the first two or three years on the job. This paradigm for graduate school died in part because the Sputnik era dissipated and in part because it simply did not work very well.

[4] The Graduate Chairperson, also called the Graduate Director, is part of the administrative structure of a university mathematics department (see Appendix I for more on administrative structure). This person is in charge of seeing that the graduate program—including admissions and mentoring of the graduate students—runs smoothly. S/he works alongside the Chairperson and other members of the infrastructure to see that the Mathematics Department sticks to its agenda and accomplishes its goals. The Graduate Chairperson is usually assisted by a Graduate Committee. See Chapter 3 and Appendix I for more about these matters.

1.1. Impressions of Life after College

where the graduate students hang out, what is expected of you. Where the coffee pot is. Do the same with the Department Chairperson.

Make a point of getting to know some of your classmates (your peers) and also some of the graduate students who are ahead of you in the program. The latter bunch will be full of a lot of gossip and a lot of baloney, but they also have passed the quals and they are familiar with how the program works. They know which classes to take, and the various hoops that a graduate student must jump through. Pick their brains. They can tell you which quals are hard and which are easy. Who writes the quals and who grades them. How quickly you are expected to get through the qualifying exams. They will know who the good instructors are, who gives good courses, and who the good thesis advisors are. This is vital information that you must know and understand.

Some of the advanced graduate students will already be writing their theses. They will have a good sense of who the good mentors are, who is hot (i.e., publishes a lot of excellent papers in top journals and gets a lot of invitations and grants), and who the "nice guys" are (among thesis advisors). [It is actually possible for a professor to be both "hot" and "nice", but don't depend on it.] It is a matter of some interest to know which thesis advisors are good at getting their students jobs and which are not. Which advisors are well-connected in the profession? Some advisors will have you going to conferences and meeting other mathematicians around the country and the world early on in your career. Others will not. Some advisors are just full of ideas and good problems; others are not. An extremely eminent mathematician at the University of Chicago once said (when asked about a problem for a student), "If I knew any good problems I would work on them myself." Perhaps he was at a stage in his career where he would not have made a good thesis advisor.

The graduate students who have been around for a while can tell you what the TA duties are like, which professors are good to TA for and which are not, which courses are a pleasure to teach and which are not. They will tell you what the undergraduates at this institution are like, whether there are discipline problems, whether the math department trains its TAs adequately and backs them up when necessary.

Of course the graduate students who have been there for a while can also tell you the good places to get lunch, have a cappuccino, see a movie, hear some music, find an apartment, and all the rest. A good way to make friends is to just hang out, listen, and ask a few questions.

These are all preliminary remarks to get you thinking about what goes on in a math department graduate program. I will treat most of these topics in systematic detail in the ensuing sections. Beginning in the next chapter,

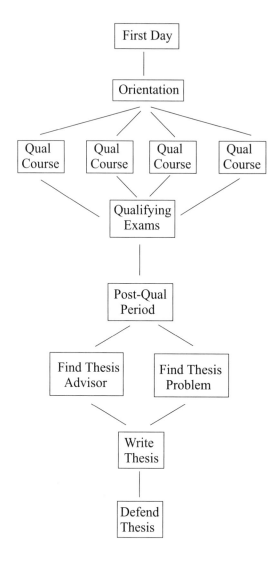

Figure 1. Steps to a Graduate Education

I will take you from being an undergraduate thinking about graduate school on to the application process, admissions, beginning a graduate program, taking the quals, finding a thesis advisor, writing a thesis, and then getting a job. We will proceed step-by-step, in chronological order. The purpose of the present chapter has been to get your mind moving in the right direction, beginning to formulate some queries.

I want you to be considering all these matters, forming questions in your mind, seeking answers. You will get more out of this book if you have some idea of what you are looking for. Figure 1 gives a sketch of the steps to a graduate education. The rest of this book will fill in the details.

1.2. What to Look For in the Future

The great thing about adult life is that you get to make choices. If you are smart and lucky, then you will make more right choices than wrong ones. If you are reading this book, then I am assuming that you are about to choose to go to graduate school in mathematics. This is not the same as choosing to buy a Chevrolet rather than a Ford. For one thing, graduate school is a long process with many complicated steps. The reward at the end is to go into a job market that is tough and competitive. If you win, then you get a good job and then tenure in an academic math department and you get to live a pleasant and rewarding life. But nothing is assured and the process is a struggle. There will be disappointments and perhaps a few setbacks along the way.

I am writing this book because I want to help you make some informed choices. I want you to do what is right for yourself. Ask questions as you read. I will try to provide some answers.

Chapter 2

Preliminaries

2.1. How to Prepare for Graduate School

It helps to be a math major, but that is not essential. Physics majors, engineering majors, and other majors also go to graduate school in mathematics.[1] [Edward Witten, one of the preeminent theoretical physicists of our time, was an undergraduate history major!] The make-or-break criterion for admission to graduate school in mathematics is this:

> Can this person get through the qualifying exams in a reasonable period of time and does this person have the intellectual resources and the drive to write a good thesis and become a mathematician?

Thus the Graduate Admissions Committee, when examining your dossier, will be looking for

1. A substantial number of upper division math courses taken.

2. Courses taken in central areas of mathematics.

3. Excellent grades in the upper division math courses.

4. Objective indications of ability to perform graduate-level work. This might include having taken some graduate courses while still an undergraduate. It also might include having discussed advanced

---

[1]Some people worry about whether it is better to have a B.A. (Bachelor of Arts) or a B.S. (Bachelor of Science). Let me assure you that it does not matter one whit. Just graduate from college.

mathematical topics with some professors, or having shown particular tenacity or insight in class, or having submitted some particularly excellent or original classwork, or having written an excellent honors thesis, or having participated in seminars.

5. High scores on the Graduate Record Exam (GRE).[2]

6. Strong letters of recommendation from substantial mathematicians indicating that this candidate has the potential to succeed in a mathematics graduate program. I will discuss this matter in detail elsewhere (Section 2.5). Suffice it to say for now that you need to get to know your undergraduate professors personally. And let them get to know you personally, so that they can comment (in their letters) enthusiastically and in detail on your abilities when the time comes. These people will not only be commenting on your technical facility with mathematics, but also on your maturity, your collegiality, and your capacity for hard work.

The right (curricular) preparation for a Ph.D. in pure mathematics and an academic career has not changed in quite some time. As an undergraduate, you should (ideally) have solid courses in

- Real analysis (from a book like [RUD] or [KRA4])
- Complex analysis (from a book like [CHU])
- Abstract algebra (from a book like [HER])
- Geometry (from a book like [DOC1] or [ONE])
- Topology (from a book like [MUN1])
- Manifold theory (from a book like [MUN2] or [LOS]).

This list is a bit ambitious and does not apply to all of us. If you are missing one or two of these, then you still probably qualify for a good graduate program. Surely you have had six upper division courses in mathematics, but perhaps you pursued logic or numerical analysis or dynamical systems instead of the traditional pure math track. That is fine. The most important thing is that you have demonstrated proficiency with advanced mathematics and an ability to think like a mathematician.

If you choose to attend a graduate program at a second-tier department, then the requirements will likely be a bit different. Such a program probably will not expect that you have thoroughly covered all six of the topics listed

---

[2]The Graduate Record Exam (GRE), administered by the Educational Testing Service of Princeton, New Jersey, is a standardized screening exam to determine eligibility for graduate school. See Section 2.4 for more details. A great many of the graduate programs in this country use the GRE as a barometer for admissions. An exam such as this clearly cannot recognize future Fields Medalists or even future mathematicians, but it can determine a basic level of proficiency.

above; but they will want you to have covered perhaps half of them, and have had some exposure to the others.

If your intention is not to pursue pure mathematics but instead to study applied mathematics, then perhaps your undergraduate preparation would be a bit different. Instead of topology you might take control theory. Instead of manifold theory you might take numerical analysis or combinatorics. Instead of abstract algebra you might take a course in mathematical physics or computer science.

If your intention is to be an actuary or a statistician then, again, certain adjustments would be made in the curriculum. You would take courses in probability, stochastic processes, statistics, perhaps the mathematics of finance, and a little work in computer languages or algorithmic thinking.

My university has one of the better graduate programs in mathematics. If we get an application from a student at The Silver Spoon College for Fine Young Ladies and Gentlemen, together with letters saying that this is the best math student they have seen in ten years, then I have to tell you frankly that it is rather difficult for us to assess the case. We don't know any of the faculty at Silver Spoon, we don't know the curriculum at Silver Spoon, we have never had a student from Silver Spoon, and we have no idea what a math major from Silver Spoon will know.

Unfortunately, there are good colleges in this country where the abstract algebra class involves primarily chatting in a vague manner about groups and rings and the real analysis class consists of looking at pictures of the Weierstrass nowhere differentiable function and trying to feature compact sets. Life at Silver Spoon may be just like that. The students are not exposed to *actually doing mathematics*—by which I mean stating theorems and proving them, constructing counterexamples, formulating conjectures, asking good questions. A student going through such a curriculum may indeed emerge as the best student they have seen in ten years, but such a student will have no hope of success in a good graduate program.

Thus a hard-nosed graduate program of good quality will look at an application from Silver Spoon and say, "What about this Silver Spoon place? Does Silver Spoon itself have a graduate program?[3] How many students does

---

[3]Today many a bright, mathematically apt student will take calculus and even linear algebra in high school (this is quite different from 75 years ago, when it was common for people to take calculus after they had taken several more basic courses in college). Such a student will hit the ground running in college and can exhaust the offerings of a basic undergraduate program in mathematics in three or even two years. What is such a student to do during the remainder of his/her undergraduate education? If there is a graduate program in place, then the answer is easy: you begin to take graduate courses while you are still an undergraduate. If not, and if the faculty consists primarily of people trained to teach undergraduates, then this student will flounder around for lack of things to do.

Silver Spoon send to graduate school each year?[4] How extensive and how deep is the math curriculum at Silver Spoon? From whom are the letters of recommendation? What does this student know? What has s/he read? What are the GRE scores of this candidate?"

Whatever may be the shortcomings of the Graduate Record Exam (GRE), it is at least an objective measure of a student's abilities in basic undergraduate mathematics. If a student is the best in ten years from Silver Spoon, with a straight A record and on the Dean's List every semester, yet the Math GRE score is 515 (out of 990), then this is a strong indication that there is some disconnect. [Believe me, we have seen applications exactly like this.] Such a student may not know much advanced mathematics.

Let me stress that the matter of filtering those whom we admit to our graduate program is not just an exercise in arrogance. Graduate school in mathematics is a long haul and very hard work. The thesis particularly is a difficult and trying exercise, and it holds a lot of heartbreak and disappointment on the road to the Ph.D. It would be doing no kindness to admit a student who is clearly unqualified. [By the same token, it is doing nobody any favors to pass a student on the quals who clearly doesn't have what it takes to pass the quals. Qualifying exams serve a purpose, and so does the admissions process.]

There was a period of time—perhaps 30 years ago—when the Washington University Mathematics Graduate Program received a lot of applications from India. There are a great many people in India named Gupta, and we received applications from many, many Guptas. One year a clerical error was made and we sent an admissions offer to the wrong Gupta. How could we tell? This was in the days before the Internet and e-mail. We just sent out ordinary airmail letters of admission to all the chosen students and waited for their replies. The Gupta in question eventually wrote

> Joy reigns supreme in my village. Nobody in this district has ever been to school before. We have sacrificed a goat.

Clearly this Gupta was not meant for our graduate program in mathematics. We had to swallow our pride and write this person a letter proclaiming our mistake and offering our apologies. There really was no alternative. It would have been a dreadful form of cruelty to let this person into our program anyway. The admissions process does serve a useful purpose.[5]

---

[4]St. Olaf College, for example, has no graduate program, but it has a hotshot math faculty and a dynamic mathematics curriculum. It ranks number five in the nation for the number of students it sends on to graduate work in mathematics each year. Other four-year and Liberal Arts colleges, such as Santa Clara University and Macalester College and Williams College, also have a high profile in the math world.

[5]Unfortunately, there was nothing that we could do for the goat. And, by rejecting the student, we could have been missing out on the next Ramanujan.

2.2. The Undergraduate Research Experience

I do a lot of undergraduate advising. A question that frequently arises from ambitious, well-meaning students is this: In order to qualify for a good graduate program, shouldn't I do some research, and publish some papers, while I am still an undergraduate?

This question demands an answer. Certainly it has become quite trendy for *undergraduate programs* to advertise that "Your kid will be doing research, even while an undergraduate, with top academics in the field." Of course, likely as not, the kid will be washing test tubes in some Nobel Laureate's chemistry laboratory, but this all sounds very exciting and it is a great thing to tell grandma at Thanksgiving.

Mathematics is a bit different. We don't use test tubes. The hard fact of life is that mathematics is a subject that builds vertically. You are not really qualified to do serious research—the sort of research that mathematicians actually *do*—until after you pass the qualifying exams. This is something that you usually do when you are a graduate student. Nevertheless, there is value in exposing a bright undergraduate to mathematics outside of the classroom setting. Surely the ambitious young math student should see that mathematics is (ultimately) *not* something that we learn from textbooks, and is not digested as received wisdom. In fact mathematics is a *process*, an endless sequence of exploration, discovery, recovery from missteps, calculation, confusion, enlightenment, and (in the end, if you are lucky) a new theorem. We can, through undergraduate research, introduce the tenacious undergraduate to library resources, to talking to other students and to faculty about mathematics, to the general idea of living and breathing and sleeping with a problem, and to the joy of the mathematical life.

When the student who has been in an undergraduate research program gets to graduate school, s/he should *not* expect to continue the "undergraduate research activities" and transmogrify them into a thesis. Undergraduate research is a developmental step in the education process. When you get to graduate school you begin anew. You learn how mathematics is done at your new institution, you study for the quals (and, we hope, pass them), you hook up with a thesis advisor, and you get involved in the advisor's mathematical world.

I can tell you that, as a student, I was accepted by every top graduate program in the country, and I never considered doing any research as an undergraduate. The same can be said for most of my successful colleagues. I am friends with a good many Fields Medalists, and none of them engaged in undergraduate research. What they did instead as undergraduates was to study like devils to learn as much mathematics as they could. That is what you should do. Doing undergraduate research can aid in this process. It can expose you to the larger mathematical world, it can help you to understand how the research process works, it can give you exposure to working on an unsolved problem. But please understand that it is a self-contained process that will usually, and should, cease when you get to graduate school.

What you are supposed to do in graduate school is find some famous scholar who will agree to direct your thesis and then you should do anything s/he tells you to do. This famous professor will already have a well-developed, widely recognized and appreciated research program. What you want to do is become a part of *that program*. *That* is the road to success in mathematics. You want to become part of this professor's crowd, talk to this professor's colleagues, learn their argot, and contribute to their well-established and successful activities. Undergraduate research is a great thing, but don't let it get in the way of your graduate education. Graduate school is a new chapter. There are a new set of rules and a new set of values—with a new curriculum and a new faculty. Begin it afresh.

Let me conclude this discussion by reiterating some of the good features of an undergraduate research program: **(i)** it gives you an opportunity to know a faculty member well, to exchange ideas, and to grow intellectually, **(ii)** it exposes you to learning outside the structured environment of the classroom, **(iii)** it teaches you to use the library and other scholarly resources to learn new ideas and internalize them on your own, **(iv)** it teaches you independence and tenacity, and **(v)** it gives you confidence and a broader vision of what the subject is about. These are all valuable commodities, and well worth acquiring. If you do engage in undergraduate research, keep it in perspective. Realize what it is, and what its limitations are. When you go off to graduate school, don't expect it to continue.

In my own mathematics department we had recently a very gifted student who went off to one of the best graduate programs in the country. He had written two terrific papers with one of the best and most famous faculty in the department. He won a major national prize for his undergraduate research. But his graduate career was dreadful—a failure by any measure. [And let me assure you that this individual was plenty smart and plenty able. The fault for his travails was in another part of his head.] He finally squeaked through graduate school, received his Ph.D., and was by then so disgusted with mathematics that he quit and went to work in a genomics laboratory. Of course this is fine—he now has a promising and productive career—but what happened to his mathematics? I asked his thesis advisor (who in fact was his *fourth* thesis advisor—he had been fired by the first three[6]), and the answer was as follows: this young fellow had the "Putnam

---

[6]See Section 6.2 for the concept of being fired by a thesis advisor.

Exam mentality".[7] He was accustomed to getting quick results without a lot of hard work. He did not have the discipline and tenacity that is necessary for the long and protracted study that is advanced mathematics. As a result, he found himself (in graduate school) in a hostile environment. And he failed. Unfortunately, his undergraduate research experience had led him to believe that the "Putnam Exam mentality" is sufficient for a successful research career.

Every mathematics department has its own style of doing mathematics. You should not necessarily expect that your glowing undergraduate reputation will precede you at your new graduate institution. What *is* true is that you have the right study habits, the right smarts, and the right work ethic to do good graduate work. Your professors have determined that this is so, they have recommended you, and you have been accepted. But now you must prove yourself. Go off to your graduate program expecting this to be a new and rewarding educational experience. Treat it as such.

2.3. Summary of the Optimal Qualifications for Graduate School

The very best preparation for graduate school in mathematics is this: **(1)** go to a top undergraduate school with well-known faculty, **(2)** perform well in your undergraduate math courses, **(3)** take as many graduate courses as you can while you are still an undergraduate, and **(4)** make a strong impression on your professors so that you can get excellent letters of recommendation. A Graduate Admissions Committee is most comfortable admitting a student from the University of Chicago, or MIT, or Harvard with forceful letters from eminent faculty whom they know. The best possible scenario is a student who has letters (from famous faculty) that say, "I would compare this young student to David Hilbert when Hilbert was at the same stage," where Hilbert is now some famous mathematician who went to the same institution as an undergraduate. A student with letters like that, who has a solid math background with top grades (and good GRE scores, naturally), will get into any graduate program in the country.

But please understand me: What I describe in the last paragraph are sufficient but *not* necessary conditions for getting into a good graduate school.

---

[7]The Putnam Exam is a competitive mathematics competition which has been held in this country since 1938. It is sponsored by the same organization that runs the Putnam Investment Fund. The exam runs for six hours on the first Saturday in December—three hours in the morning and three hours in the afternoon. There are six problems in each of the two sessions. These problems are elementary but tricky. Those who excel at the Putnam Exam become rather famous. The first-prize winner receives a fellowship to Harvard, and the others receive substantial cash prizes. Those schools (including my own) that have a record of top performances on the Putnam are held in high regard.

I went to the University of California at Santa Cruz—never a mathematical powerhouse—and I still got into all of the best programs in the country. I wasn't the shoo-in that I would have been had I graduated *magna cum laude* from MIT, but I did just fine. The most important thing is to excel in a solid undergraduate math program and to have a transcript and letters that validate your success.

One of the great things about life in this country is that we can chart our own path. Certainly the range and variety of colleges in the United States is remarkable. There are colleges which are also working ranches, there are colleges that will train you to be a hosteler, there are colleges where you can get a scholarship in rodeo, there are colleges where you can become a technical consultant while you are getting an education, and there are colleges that will turn you into a performing artist. You are in effect shaping your life by the type of college that you choose. But a definite corollary of this observation is that the choice you make can also limit what you will be able to do in the future.

If you choose to go to a small, private, evangelical college where the emphasis is more on Christian missionary work and less on academics, then you are going to have a heck of a time getting into a good graduate program in mathematics. The Graduate Committee at a good university will be unfamiliar with your college's curriculum and with its faculty. It will not perceive you as a good risk. As noted above, I was an undergraduate at the University of California at Santa Cruz (UCSC). This was in the late 1960's, when UCSC was one of the hottest undergraduate institutions around. It was selective, innovative, and exciting. One of the novelties was that, at UCSC, there were *no grades*. I can tell you that Princeton and Harvard and MIT were quite skeptical of admitting me. They did not know what to make of this upstart institution, of the lack of grades, or of the unfamiliar curriculum. In the end, because my record was so outstanding and my recommenders well known, they decided to take a chance on me. After I did well in graduate school, the door was opened for future students from UCSC. But it was a battle for me to get admitted that first time.

2.4. What About Those GRE's?

The GRE Test (or Graduate Record Exam) is a part of life these days. It is a standardized test, administered by the Educational Testing Service in Princeton, New Jersey, to measure proficiency and preparedness for graduate school.[8] Most graduate programs request that you take the GRE as part of the application process.

---

[8]This is the same organization that writes and administers the Scholastic Aptitude Test, or SAT (to screen high school students for admission to college).

2.4. What About Those GRE's?

This is not to say that they attach a huge amount of significance to the GRE. Certainly your course grades, the range and depth of courses you have taken, and your letters of recommendation are the most significant features of your graduate school application dossier. But the GRE is the great equalizer. It is an objective measure of whether you know anything. If you come from a school or a program that is unfamiliar to the Graduate Committee at the school to which you are applying, or if your background is unusual and difficult to put into context, then the GRE may play a significant role in measuring your qualifications for graduate school.

The GRE has two parts: **(i)** the general portion that is divided into "verbal skills", "quantitative skills", and "analytical skills" and **(ii)** the advanced subject-area portion (in your case this will probably be the advanced mathematics test).[9] As part of your graduate school application procedure, you will take *both* parts of the GRE exam.

Each of the three parts of the basic portion of the GRE is worth 800 points. A perfect score on the advanced mathematics portion of the GRE is 990 points (rescaled recently from 800 to prevent clustering of grades at the high end). A person intending to go to graduate school in mathematics certainly should score *at least* in the high 700's on the advanced mathematics subject area exam. As I have said elsewhere, if your score is instead in the 500's (or lower), then many eyebrows are going to be raised. The GRE is a pretty good, if not a profound, barometer of your general mathematical abilities. If you cannot score well on the GRE, then perhaps you have chosen the wrong career path.

Can you prepare for the (advanced math) GRE? Contrary to what the Kaplan Service, the Princeton Service, and the other pay-as-you-go test preparation agencies (which offer prep for the general portion of the test, not the advanced subject-area portion) want you to believe, the answer is that I think not. You can, if you wish, prepare for the basic GRE by spending a couple of days with one of the standard prep books, familiarizing yourself with the type of questions that appear on the GRE and the standard subject areas that are covered. If your undergraduate education has obvious gaps then perhaps you really will have to study. But the basic test is measuring what you've got in your gut—not what you've packed into your cranium at the last minute.

The GRE Math Test examines for basic knowledge and facility with advanced mathematics. If you have a solid background, and if you get a good night's rest before the exam, then you should do fine. It is quite

---

[9]The advanced mathematics subject test consists of 66 multiple-choice questions. They are broken down as: 50% calculus, 25% algebra, and 25% other topics. Further information is available at the GRE web site: `http://www.gre.com`. You can also learn about the grading of the exam from the GRE booklet `ftp://ftp.ets.org/pub/gre/01210.pdf`.

an ordeal to take both the general portion of the GRE and the advanced subject-area exam on the same day. It might be wise to schedule them on different days. But, to come to the point, I can see no reason to study for the Math GRE, and I do not encourage you to do so.

2.5. How to Choose and Apply to a Graduate School

After you were in college a few years, you surely realized (either consciously or unconsciously) that reading the college catalog gave you only the vaguest of ideas of what this institution, and this educational experience, would really be like. Reading college guides that you bought at Borders (or reading the copious information that is readily available online) perhaps gave you a slightly different perspective,[10] but again no hint as to how your college education was actually going to play out.

In fact the situation is more complex for graduate programs. No catalog will tell you that Harvard has very little analysis, that Stanford has very little algebra, that there are only a few schools with a real program in mathematical logic, that the Courant Institute is the preeminent applied math program in the country, that the Princeton math program has no basic (qual-level) classes and no grades, that at some schools most professors will refuse to direct a thesis, or that at other schools most of the finishing Ph.D. students cannot find a job (at least a job in academics). *You must talk to people, mainly your professors, to really find out what the different graduate programs are like and which program is best suited to you and your abilities.* The official college catalogs are going to give you sugar and spice and everything nice. Some of this may actually be true; much of it will be fanciful. You get the truth by talking to people.

It is OK to talk to your fellow undergraduate students, but they do not know much. How could they? We are talking about high-level professional preparation for an elite career. You had better get advice from people who have hands-on knowledge of that career path. You really must talk to the mathematics faculty at your undergraduate institution. To begin, certainly discuss the matter with your undergraduate advisor. Your advisor will have an excellent idea of your abilities, of which programs best suit your interests, and s/he can chat with you about where the nice campuses and pleasant environments are.[11] Some excellent universities are located in

---

[10] For example, *Lisa Birnbach's College Book* [BIR] is a hip tour of American colleges. It will tell you the favorite drugs on campus, how to get laid, and offer other unconventional wisdom about campus life.

[11] My view is that graduate school is tough and demanding. You are going to be in it for a considerable period of time. You may as well live in a place where you feel comfortable, where you can find relaxing diversions and nice places to eat—where you can go to a symphony or a ball game or a race track if you wish. Fortunately, this country has many good graduate programs so

2.5. How to Choose a Graduate School

dreadful parts of big cities. Others are quite isolated, and travel to and from them is very difficult. Some universities are virtually inaccessible during the winter. Some good universities are on the beach; others are in the mountains. Some have very friendly faculty; some don't. Some get their Ph.D. students through the program in four or five years. Others have students that languish in the Ph.D. program for ten or more years. In some programs most of the students never finish, and this is what the math department intends (because the graduate students' primary function is to teach recitation sections and not so much attention is paid to their progress toward the Ph.D.[12]). You are not going to get this information by going to a web site or buying a book at Borders or reading a catalog. You've got to *talk to people*. A good and experienced faculty member will actually have been to many of the universities, will know some of the faculty there, and can really tell you chapter and verse on programs, professors, and curricula.

One common concern centers around the *size* of an educational program or institution. One hears these concerns about size voiced most stridently among students applying to undergraduate school. The common lament is "This school is large and impersonal. I will be lost, and not get the mentoring and attention that I need." By analogy, one can imagine a person saying, "I don't want to live in the United States because it has nearly 300 million people. I'm going to move to Liechtenstein."

There is a serious point to be considered here. What is good about a large university, or a large graduate program, is that you (the student) have many choices. There are a great variety of faculty, and a great many areas are covered in depth. There will be many specialized subprograms and interdisciplinary programs. [By contrast, Harvard, which has one of the highest ranked graduate programs in mathematics, has only about fifteen permanent faculty and concentrates on just a few particular areas of mathematics.] Living in a large university or a large graduate program is indeed like living in the United States. What you do as a resident of the U.S. is you choose a state, and you choose a town, and you choose a community in which to live. *Those* are the people whom you see on a daily basis, and with whom you make your life. Just so, in a large academic program you choose an area of study, and perhaps even a concentration within that area,

---

you have considerable choice. Indeed, there are 48 Ph.D. programs at the Group I universities, 56 at the Group II, and 72 at the Group III—see the Glossary for more information about these groupings.

[12]In fact there have been some programs that were well-known for admitting a number n_1 of students, with the conscious idea in mind that some number $n_1 - n_2$ would never make it past the quals. The program—in the long run—could really only accommodate n_2 students. But they needed a flow of warm bodies to act as TAs so that their calculus classes could be taught. It is really a dreadful system, and I am happy to say that public attention has caused this exploitative setup to begin to fall by the wayside.

and then you develop a relationship with a couple of professors and a couple of graduate students, and that is how you make your life. The down side of the large program is that, during the first couple of years when you are trying to find your way and to get through the quals, you really can become lost. You can want for advice and companionship. You can lose your vision and your drive. And, sad to say, many students do.

By the same token—and to repeat—the disadvantage of a small college or program is that there are fewer choices. If you are in a small department where geometry and logic are the strengths, and if you decide after a few years that what you really want to do is analysis, then you are stuck. Or you could find that a small department has a rather stuffy and unwelcoming atmosphere that does not suit you; if there are only fifteen people on the faculty and you cannot see eye-to-eye with any of them, then you have nowhere to turn. Again, the best way to get a feeling for what different departments are like is to talk to the faculty at your undergraduate institution. Like myself, they will have been around for a while, they will know many of the departments and many of the faculty, and they can give you a feeling for what you are getting into. Certainly one consideration to keep in mind is this: if you have no idea what area of mathematics you want to study, then a large department will offer you the broad array of choices that you may need.

You can get copious information (of a formal, packaged nature) about any math graduate program by sending a postcard to

> Graduate Director
> Department of Mathematics
> Valhalla University
> La La Land, Missouri 63144

An *e*-mail will also often do the trick. Many math departments have a detailed web page with a lot of good information and with a form that you can fill out for additional, hard-copy information. And many allow you to apply online.

You can go online and view the departmental web page for any math program that strikes your fancy. Read about the faculty and their areas of interest. Have a look at some of their recent publications. The AMS publishes lists of recent Ph.D.'s each spring. Look at the theses titles and see what kinds of mathematics come out of the school that you are considering.

I usually advise students to apply to two or three dream schools, two or three realistic schools, and two or three fallback (or "safety") schools. Berkeley, Princeton, MIT, and Harvard have led the ratings for many years. Yale, Chicago, Michigan, UCLA, Duke, and Cornell are also top-rated. There are

2.5. How to Choose a Graduate School

many other schools with excellent programs. The National Research Council (NRC) periodically (about every ten years) ranks graduate programs in all academic disciplines.[13] The *Notices of the American Mathematical Society* reports on the NRC mathematics rankings (see particularly the articles on the 1982 and 1995 rankings and the new ranking that is being prepared as of this writing); these articles are widely read and debated.

Part of the NRC ranking is that each math department is rated as Group I, Group II, ..., up to Group V.[14] Schools take great pride, and attach real significance to, what group they are in. Group I (48 schools in 1995) comprises the very best schools. Group II are solid schools (56 institutions in 1995). Group III tend to be more modest, regional schools. The production of Ph.D.'s at a Group III school is often low and episodic. Groups IV and V are not part of the linear pecking order. Instead, Group IV designates programs in statistics, biostatistics, and biometrics; Group V designates applied mathematics and applied science.[15]

It costs about $40–$50 per school to submit an application, though some schools will waive the fee if you apply online (using the internet). If you are tempted to apply to 20 graduate programs—just to give yourself a choice—then it is going to run you into some money. Usually, applying to about eight schools is more than sufficient to cover all of your options.

Most schools have an application deadline in December or January (for the following fall). The main thing is that *your application* needs to be submitted by that deadline so that they can open a file on you. Your letters of recommendation (usually three) should arrive shortly thereafter. There is no penalty against those whose applications arrive late—*except* that the Graduate Committee will begin screening applications, and if yours is too late, then all of the positions and support may be gone.

Most everyone who applies to graduate school in math has good grades (at least in their math courses) and good scores on the GRE Test. So the make-or-break factor in your dossier will be your letters of recommendation. Take care in choosing whom you will ask for a letter. It should be someone who knows you well and who *knows your mathematical abilities*.

---

[13] As with many rankings by "objective" agencies, this one can play a significant role in university life. Deans, for instance, pay particular attention to such an ordering. The NRC ranking is heavily dependent on self-reported data (which is sometimes inaccurate). A large component of the ranking is overall reputation, or memory of a favorable graduate experience a long time ago.

[14] All of the details about the NRC ranking of graduate programs may be found at http://www.utdallas.edu/~sudip/nrc.htm and also in the articles [JAC], [MAC], [NRC], and [GMF].

[15] Another important ranking of graduate programs is the Carnegie Classification, established and administered by the Carnegie Foundation. An internet device for creating your own ranking may be found at the Courtright Memorial Library whose web address is http://www.otterbein.edu/resources/library/libpages/subject/subgrad.htm.

Letters from your priest or your cub scout leader or your mother are all fine and well, but they don't do much toward getting you into graduate school. Mathematicians trust letters from people whom they know and who speak their language. I got into Princeton (my dream school) in part because I was an excellent student and in part because I made sure that all my letter-writers were well-known at Princeton. A letter from a physics professor or a computer science professor is fine, but most of your letters should be from mathematicians. I have gone into some detail, in another venue, about what makes a good letter (see [KRA2]), but this is not your concern. You will not be writing the letters.[16]

You want letters that will speak to your specific mathematical attributes: Did you ask interesting questions in class? Did you show some special qualities? Did you evince originality? Tenacity? Great technique? Are you well-read? Did you ace all of your exams? Construct brilliant and original proofs on the fly? Write an impressive honors thesis? Did you do well on the Putnam Exam? Any of these might make a good impression on the Graduate Committee at your chosen university. It is also helpful to the Graduate Committee if your letters of recommendation speak to your maturity, your capacity for hard work, your congeniality, and your overall balance and good sense.

As previously indicated, three is the standard number of letters to obtain (each school will tell you how many it wants). Do *not* be bashful about asking for letters. This is part of a professor's job. S/he is supposed to write letters for students. Very occasionally a professor will tell you that s/he doesn't have time to write, or doesn't feel that s/he could write a good or supportive letter about you. Don't worry about it. Just ask someone else. My recommendation is that you ask for the letters two months before they are due. Go back and remind the professors in a month. At that time you should phone or *e*-mail each program to which you have applied and ask whether your dossier is complete.[17] Then you will know who the laggards really are. [Many schools will help you out by sending you a postcard or *e*-mail notifying you whether your application is complete.] Remind the professors again after another two weeks.

---

[16]Unfortunately, some students aspiring to graduate school figure that they cannot get solid letters from their professors, so they *forge* their letters of recommendation—actually signing the names of famous faculty. This is a big mistake. Academic mathematics is a small world, and the student's likelihood of getting caught is quite high. First of all, most faculty can recognize a letter from virtually any well-known mathematician without even looking at the signature. Secondly, many schools now send out an acknowledgement for a letter of recommendation—directly to the professor who is alleged to have written it—specifically to discourage the subterfuge described here.

[17]A friend of mine in college did not get into Harvard Medical School—and he probably would have, as he was *that good*—because Harvard lost his application.

Even though most schools will ask for three letters, it is OK to request letters of four or five people. Don't overdo it. Fifteen letters is inappropriate. A copy of your scrapbook is inappropriate. A compendium of your Eagle Scout Merit Badge Citations is not helpful. A videotape of you teaching is out of place. But a couple of extra letters for insurance is fine.

Another interesting feature of the application process to which many applicants give too little attention is the essay, also called the Statement of Purpose. You are usually asked to write a page or so on why you want to get a Ph.D. in mathematics. *Do not blow this off.* Many programs, especially the best ones, get a great many students with virtually isomorphic credentials (straight A's, perfect scores on the GRE, rave letters, etc.). They need some device for differentiating among students, for determining who's got what it takes.[18] This little essay is certainly one way for them to do so. You should really give some thought to your essay. Write a draft and show it to some people. Revise it. Let it sit for a while and go back to it. *Show it to your advisor.* Your best friend. Make it as polished and compelling as you can.

Do *not* write a sappy or maudlin application essay. An essay that reads

> Mathematics is the most sublime and poetic subject in the universe. Its rapturous elegance and intrinsic harmony make my life worth living, and I evince mathematics with each breath that I take. It would be the honor of my life to earn a Ph.D. in mathematics and to carry forth the eternal torch of learning into the future for succeeding generations of young scholars ...

will not impress anyone, and is liable to turn away many. Just tell the truth about why you want to be a mathematician, what you like about the subject, and what your career plans are. Emphasize features of your background that may be relevant to your application but not apparent from other parts of your dossier.

Make every effort to get your application in punctually. Be sure that it is addressed correctly and has plenty of postage. Don't forget to enclose your check (for the fee). Be sure that your contact information (mailing address, *e*-mail address, FAX number, phone number) is included.

Most graduate programs have an agreed-upon date when they will let students know whether they have been admitted. This information should

---

[18]Princeton University has one of the top math Ph.D. programs in the world. All of the best students apply there, and they have sensational credentials. It really is quite difficult to tell them apart. For many years, Professor Ralph Fox exercised an uncanny knack for reading the essays and telling which students had the necessary drive and promise. His special ability became known as the "Fox sniff". Professor Fox is now deceased, and I have instead had Princeton faculty phone me for extra insights on applicants (from my university) to the Princeton graduate program.

be in your application materials. Then you will also have a designated period of time to think about the matter and render a decision. Some departments have the money and discretion to fly in graduate candidates for a visit (after they have been accepted, but before they must respond). If you are invited to do so, then I encourage you to go. Armed with some of the questions that I have raised here (see also Appendix IV), you can learn a great deal from a first-hand visit. You will meet some graduate students and faculty and you can ask some hard (but polite, please) questions. Then you can make a well-informed decision.[19]

These days, a number of students finish their baccalaureates in December (rather than May or June), so they wonder whether they can begin graduate school in mid-year. The answer is "yes", it is theoretically possible, but your choices will be few. Most graduate programs want to develop the *esprit de corps* of each class and they want students to enter as a group. More importantly, most of the basic (qual-preparation) courses are year-long courses that begin in the fall and most qualifying exams are given in the spring (May or June). So a student who begins in mid-year will be out of sync with all of the most important activities for beginning students. There also may be no financial support to offer a mid-year student. So most graduate programs in mathematics discourage mid-year students.

In short, you will be at a distinct disadvantage to begin graduate school in mid-year, you will only be able to choose from a limited number of schools, and I advise against it. If you finish your undergraduate work in mid-year, then I suggest you try to get a job in a lab, in industry, or as a research assistant to a professor. Or do as I did—just enroll for an extra semester of school so that you will finish with everyone else. You can use that extra semester to take a reading course in an advanced subject area, to burrow more deeply into a subject that you've only studied cursorily, or to relax and take a literature course.

In Europe it is fairly common for a student to study for a while at University A and then to move to University B (possibly even in another country!) where there is a particular professor working on things that are of particular interest. The funding situation is set up in such a way as to make this mobility both feasible and natural. Such horizontal movement is virtually unheard of in this country.[20] For the most part, you will pick a

---

[19]I can tell you that most of the students whom we fly in are quite impressed by all of the attention they receive. They get to meet students, faculty, staff, and even some administrators. Efforts are made to show them the city and acquaint them with life at Washington University and in St. Louis. This is a great recruiting tool for us.

[20]In Boston, New York City, or Los Angeles—where there is a high density of excellent universities—one does sometimes see cross-fertilization among universities. A student at Harvard will end up working with someone from MIT or a student at NYU will end up talking to people at Columbia, but the incidence of this type of activity is quite limited.

graduate program and spend five or more years in it. You will not move and you must grow where you are planted. This is a significant segment of your life and of your education. It will have a huge impact on your future, so you must choose wisely and carefully.

2.6. Special Considerations for Underrepresented Groups

Sad to say, women, African-Americans, Native Americans, and some other groups are underrepresented in the mathematics profession. That is to say, the percentage of women in mathematics is markedly lower than the percentage of women in the general populace, and likewise for the other groups. This fact is a product of insidious societal influences that follow from the special nature of America's history; they are beyond the control of most of us. But we try to learn from our past mistakes; today we strive to make our mathematics departments as welcoming and nurturing as possible for all people.

If a woman enters a graduate program in which she is the *only woman*, then she is going to feel quite isolated and unsupported. The same can be said for other members of underrepresented groups. Thus, such a person shopping around for a graduate school will want to see that there is an available peer group and cognate support services. Are there special fellowships for underrepresented groups? Special mentors and counselors? Special housing? Are there women or African-Americans or Native Americans on the faculty?

In fact, this last point is worth dwelling on. As we develop, both personally and professionally, we all seek role models. Of course the person who becomes your mentor and role model will do so in large part because of special traits—particular brilliance, inspiration, a remarkable work ethic, or a strong moral code—but it is also the case that personal features can and will play a role. People feel more comfortable with others who are perceived to be "like themselves" in key ways. Especially if a student is having trouble, or is plagued by self-doubt, the student will seek out like people with similar backgrounds. Thus it is important for our graduate programs to be able to offer rich and diverse faculties.

Information about faculty diversity, and programmatic diversity, is readily available—both in print and on the web. If you have any doubts or any questions, then do not hesitate to pick up the phone and just ask. You want to be comfortable and happy and to feel that you are supported and valued in your new environment—in which you are going to be spending several key years of your life. So endeavor to make a well-informed decision.

2.7. What About My English?

This is not a trivial question. In most mathematics graduate programs you will have to teach as part of the package deal that creates funds for your tuition and living stipend. These days, math departments (and all other departments) are being held accountable for quality teaching—see Sections 3.11 and 3.12 as well as the book [KRA1]. Gone are the days when we could take a student from China or Russia or Yugoslavia—with only marginal language ability and no teaching experience—and stick this ill-prepared TA in front of a class. We have to screen our people carefully. At the *least*, most every graduate program will require international applicants to take the TOEFL (Test of English as a Foreign Language). This is a standardized and universally recognized exam to screen students for minimum ability with the English language. If you cannot pass the TOEFL, then you will likely not get into graduate school in this country.

Of course it is one thing to take and pass a standardized written test for basic vocabulary and usage and quite another to be able to function in front of a classroom—to give a lecture, to answer student questions, to interact profitably and effectively with the students.[21] At my own university, we first put each graduate student through a semester-long "Teaching Seminar" on how to teach. No student can teach in the Washington University Mathematics Department before taking and passing the Teaching Seminar. Then we also make each student give a practice lecture before a critical audience—the Graduate Chairperson, the Head of the Teaching Center, the Undergraduate Chairperson, and a few others. If the student's English is not up to snuff, then the prospective TA is told to take an English as a Second Language Course. If that does not remedy the problem, then (at the *least*) the student is relegated to second-class status. S/he is given less desirable duties, more of them, and is paid less. Of course such a student will have no teaching experience to show when applying for a first job; if the student is foreign to boot, then that student is going to have a very tough time on the market.[22]

Many international students who come to graduate school in the United States end up wanting to stay here and to develop careers here. Good for

---

[21] We in fact send an emissary to China each year—the Graduate Dean funds the trip—to interview our Chinese candidates for the Ph.D. program. We can tell a lot more about candidates by talking to them in person than by reading their TOEFL scores.

[22] I am sorry to say that, in the 1970's and 1980's, we mathematicians as a profession were a bit disingenuous. In our letters of recommendation we brashly asserted that all of our students were great teachers and charming conversationalists, even when it was plainly not true. Many a department chair was chagrined to meet a new faculty member who had only marginal English and no idea how to teach. So, having been burned many times, departments today are very cautious.

them. But if they want to get jobs in this country, then *they must be able to teach, and it must be demonstrably true that they are very good teachers.*

Unfortunately, American undergraduates these days are rather parochial. They react negatively to a teacher with a heavy foreign accent—especially one who is marginally incomprehensible. If the teacher is really extraordinary and has a charming (yet comprehensible) foreign lilt, then students will learn to recognize the qualities of the instruction and also to enjoy the value added by an infusion of foreign culture; but if the instructor is *really* difficult to understand, then the prospective mathematician will be a failure as a teacher and his/her promising mathematical career will die on the vine.

I cannot over-emphasize this last point, even though I may appear to be getting tiresome. More and more, in the past decade, I have seen *superb* graduate students from top universities *forced* to abandon an academic career just because they could not teach; in many cases, this was because of their English. These people had tremendously promising mathematical careers (measured by the yardstick of research ability), but they could not effectively perform that special function by which Deans and administrators measure our worth: they could not teach. End of story.

Most universities have considerable resources to help you bring your English up to speed.[23] There are also resources to help you with your teaching.[24] But, as usual, the onus is on you to expend the effort to make it happen. Make a point of hanging out with *American students* and speaking English. If you are from China (just as an instance), then it is easy and natural to spend your leisure time hanging out with other Chinese—we all feel most comfortable speaking our native language with people like ourselves. But you are not going to learn any English that way. You must take special pains to expose yourself to American English and to the American way of life in order to acclimate yourself to the language and cultural modes of this country. If you are going to function effectively as a mathematics graduate student and professor here, then you must take responsibility for fitting in.

2.8. How to Pay for Graduate School

When I told my parents that I had been admitted to Princeton University for graduate study they nearly suffered myocardial infarction. It had been all they could do, with their limited financial resources, to put me through

---

[23]What you are looking for here is the "English as a Second Language", or ESL, Program. The teachers in the ESL programs are talented people who work with students individually to help them with pronunciation, vocabulary, and other communications skills. I urge you to take advantage of this resource, as appropriate, at your institution.

[24]For instance, many schools have a Teaching Center for just this purpose.

college. How in the world were they now to muster the pecuniary assets to send me to an Ivy League school for five or more years?

Of course what they didn't realize, and what you may not realize, is that there is a considerable amount of subsidy in this country for graduate study in the sciences. Most graduate students, at most universities, receive full financial support. This includes tuition plus a stipend for food, rent, and incidental expenses. You won't live like a potentate, but you will be able to afford an apartment and a modest vehicle and you will be able to eat dinner out once per month if you don't order an appetizer.

The other bit of good news is that, unlike support for undergraduate study, the support for graduate work is *not* based on financial need. You do not need to fill out any long and painful financial declarations. Often, in fact, all you do is check a box on the application form to indicate that you want to be considered for financial support.

The standard methods of support for graduate students are:

TA-ships: The Teaching Assistantship (or TA-ship) is a work/study arrangement by which the graduate student performs certain teaching duties for the department in exchange for full support in the graduate program. Usually the department covers the student's tuition and also pays a living stipend. At some schools, there will be extra money for books, for travel, or perhaps other perks. See Sections 3.11 and 3.12 for further details about TA-ships.

RA-ships: The Research Assistantship (or RA-ship) is a work/study arrangement by which the graduate student assists one or more professors in their research in exchange for full support in the graduate program. See Section 3.14 for further details.

Fellowships: A fellowship is like a scholarship: your studies are supported, but you are not required to perform any service in return. Most graduate programs at the Group I schools, and many at the Group II schools, have at least some fellowships, and they are used to attract the best students to the program. There are also fellowships available from the National Science Foundation, the Department of Education, and from a variety of charitable foundations (the Guggenheim Foundation, for example, provides thesis-preparation fellowships). Members of special groups, especially underrepresented groups, may be eligible for particular fellowships.

Tuition Remission: Some graduate programs will offer to pay some or all of your tuition (or, more precisely, to waive a portion of your tuition) but will not offer any living stipend. This circumstance may be most likely to occur in a Master's program.

2.8. How to Pay for Graduate School

Most mathematics graduate programs have a sizable budget (from the home university) for the support of graduate students. There are other sources of financial aid as well. These include the National Science Foundation, the Department of Education, the Department of Defense, the Hertz Foundation, the Exxon Foundation, and a number of other private foundations. Many professors have research grants with line items for the support of graduate students. Your undergraduate college will have an office, as well as library resources, to help you in your quest for financial aid for graduate school. There is a lot of information on the web. If you have the talent and drive to be a mathematician, you should *not* have to pay for your graduate education.

I must stress, however, that there is only a finite amount of money in the world. Priority for funding goes to Ph.D. programs over Master's programs. It is more difficult to get financial aid for a Master's degree.

Don't turn the selection of a graduate program into a bidding war (in your head). So what if school A is offering $300 more per year, or school B has parking that is $12 cheaper per month? There are much more substantial considerations than these. Try to keep your priorities straight!

Some graduate programs can offer you financial support during the summer. This can really make a difference, as it may mean you don't have to work and instead have three or four months for unrestricted study. Be sure to inquire about this possibility.

What about health insurance? The graduate student stipend is, after all, rather modest. And, unfortunately, these days you must pay income tax on that income. Health insurance is expensive. How can you manage to protect yourself? The answer is that almost all university graduate programs provide subsidized health insurance—both for you and for your family. You may have to pay a modest fee, but the coverage is good, often comparable to the coverage that faculty and staff at the university are provided. At my own university, the health insurance even covers students during the summer and when they travel. So, generally speaking, health coverage is not a matter for concern.

Part 2

Essential Elements of a Graduate Education

Chapter 3

Pre-Thesis Work

3.1. Your Early Courses and Preparation for the Quals

For the first two years or so of graduate school you will be pre-thesis, and your main goal is to get through the qualifying exams (Section 3.3). If you are lucky, then your graduate program will provide you with a preliminary advisor (the graduate program where I was a student called this person your "Part Zero Advisor") to help with this process. If it does not (and woe is you if that is so), then knock on the door of a friendly faculty member and just ask. If you goof up this process, then you will not be prepared for the quals, you will not pass the quals, and you will lose valuable time.

Let me record an important observation once and for all: You should try to get through your qualifying exams by the end of your second year in graduate school. The end of the third year is OK; the end of the first year is just dandy, but two years is the usual goal. One of the first things you should find out when you arrive at your new graduate school is the usual paradigm/timeline for the quals; do your level best to stick to it. See Section 3.4 for a more discursive discussion of when to take the qualifying exams.

Some graduate programs (such as the one here at Washington University and also the one in which I was a graduate student) have a well-defined collection of quals (qualifying exams) that every graduate student must pass. At my present institution, there are four qualifying exams (in algebra, geometry/topology, complex analysis, and real analysis) and a year-long course to go with each one. You take each year-long course and then take the qual at the end (written and graded by the professor who taught the course). Thus the curriculum for a beginning student is easy. In the first year, you pick two year-long graduate/qual courses plus one other course (for a total of nine credit hours) each semester, and you take two quals at the end of the year. In your second year you do just the same. After that you start looking for a thesis area and advisor (Sections 4.1 and 4.3).

At other schools things are more flexible (or less structured). There may be more or fewer quals and they may take up a different portion of your graduate career. I used to teach at Penn State, and they had (at least in those days) a broad menu of qualifying exams. The student was to pass a certain number of them (three or four) and s/he could choose which ones to take. It was typical for an applied math student, let us say, to pick real analysis, differential equations, applied math, and numerical analysis as the four qualifying exam subject areas. It was typical for an algebra student to choose abstract algebra, number theory, logic, and category theory for the four qualifying exam subject areas. Penn State had this system because the powers that be wanted their graduate students to get rapidly involved in advanced work in the thesis area. The faculty did not want students to be distracted by ancillary subject matter.

Some graduate programs have distribution requirements (analogous to those that may be familiar to you from your undergraduate days). For example, your graduate program may *require*—independent of the qualifying exams—that you take a certain number of courses in algebra, a certain number in analysis, and a certain number in geometry. On the one hand, these strictures can be good for the broader goals of your education. On the other hand, they can act as a stumbling block toward getting through those qualifying exams and getting on with your life.

There are a variety of paradigms for the qualifying exams in a graduate program and you should examine this matter carefully as you select a graduate school. If the qualifying exam process is discouragingly complex or protracted, or if there is not a direct connection between the preparation process for quals and the quals themselves, then you may wish to seek a more attractive or better-designed program. There are some fairly famous programs that have a quite Byzantine system consisting of *two* (not one) substantial sets of exams, or several individual eight-hour exams. These quals are quite daunting and the attrition rate (i.e., the percentage of students who do not get through the quals) is rather significant. Perhaps such a system is not for you.

The point is this: Most every graduate program has a set of quals, and a cognate set of courses to help you prepare for the quals. You don't *have* to take those preparatory courses, but that is the standard way to go. If you get good advice, then you can tailor your choice of courses and your program of study to your needs, your abilities, your background, and your aspirations. And you can get through the quals efficiently and effectively.

3.1. Preparation for the Quals

Always remember that the qualifying exams *are not the point of graduate school*. They are just a step along the way. The main thing is to write a good thesis. So your short-term goal, at the beginning of your graduate career, is just to get through those quals. The quals are a zero-one game. Once you have passed them, then you need never look back.[1] It's time to write the thesis.

If you pick a reasonable graduate program with a reasonable system of qualifying exams with which you can cope, then your coursework in the first two years should be directed to preparing for those exams. Perhaps you had a little graduate work while you were still an undergraduate; this might get you excused from some quals. Or it might mean that you do not have to take the cognate qual course at your present graduate institution; you can just sit down and take the exam. You will have to meet with an advisor or with the Director of the Graduate Program to determine whether this is so.

After you get past the quals, then the panorama of courses is a wide-open tableau for your delectation. You will still need to choose wisely. Your goal now is to find a thesis advisor and problem, and you should bear this in mind as you select your classes.

Once you have a thesis advisor and a thesis problem, then it will be natural and easy to choose appropriate classes. By then you will probably have satisfied the course requirement[2] and you can choose just those seminar-level classes that interest you and will perhaps help you in your thesis work.

In ancient times, when I was a graduate student, we only took pure mathematics courses. I would have been mortified to tell my fellow graduate students that I was taking a course in "C" programming or a course in mechanical engineering. Today, matters are different. For one thing, we all have a much broader view of mathematics and we appreciate its uses outside of the pure realm of theorem-proving. The most solid mathematicians with the most robust research programs are those who have a broad base of knowledge and experience. You should begin to build that base while still a graduate student. There are also practical reasons for taking courses outside the strict stream of mathematical thought. It will make you more attractive on the job market if you have some exposure to computing and/or to areas where mathematics is applied. If you are seeking a third course to

---

[1]To be honest, there may be times when the Graduate Committee is trying to pick students to whom to give special perks, or summer support, or a special prize. Naturally the Committee will gravitate toward students who have performed well on the quals. It is always better to excel when you can, but your score on the quals will have little effect on the writing of your thesis.

[2]Most every graduate program, be it a Ph.D. program or a Master's program, has something called a "course requirement". It demands that the student do a certain number of units of course work. This requirement is, of course, in addition to the qualifying exams and the thesis.

take, consider looking around campus to see what is going on in physics or biomedical-engineering or computer science.

One thing that you begin to realize while you are a graduate student is that learning does not have to be a formalized process. You do not need to take a course, with a teacher, homework, and a grade, in order to learn a new subject. By the time you reach an established level in the academic world, you probably will not have the patience to sit through courses; instead, you will learn things entirely on your own. An intermediary step to that lofty position is to develop the habit of *auditing* courses. One of the things we were taught right away in the Princeton graduate program is that we should think of the courses just as we think of the books in the library: these are resources that you can drop in on, and drop out of. You don't need to register. Just access them as your interests and your studies dictate.

Certainly, once you have satisfied the course requirement in your graduate program, there is no need to register for *any* courses. [You still must register for a certain number of credit hours each semester, but these can be for thesis work or for reading courses with your advisor.] And, since you are completely dedicated to writing a thesis—which is certainly a full-time job—you do not want to be bogged down with homework and being obliged to attend lectures three days per week. But you do want to keep learning new things. If some world-class scholar is giving a set of lectures on a hot new topic, you don't want to miss out. So *audit*! That means you get to attend, the lecturer can put you on the class list, and everyone is happy. You can still be flexible, you can skip class when you need to, you are not obliged to do the homework or take the exams, and you can still live a full and productive life.

I will conclude this section by enunciating a very important principle (which will be repeated often in this text) of getting an advanced education. You are no longer learning calculus or another trivial subject where it is sufficient to read the text and do the homework. You are now doing the toughest thing you will ever have done in your life. It is *essential* that you talk to people—all the time. In this way you can orient yourself, keep to your course, be sure you are doing the right thing, and have a constant reality check. It is also an important part of being a mathematician to be able to communicate—not just technical mathematics but also information *about* mathematics, about teaching, and about the profession. You are now not simply *learning* mathematics—you are learning to *create* it. So my advice is to talk to your fellow students and to the faculty (and to the staff) about *everything*. Eat lunch with a group, socialize, talk to your office mates. This is your new life.

3.2. How Am I Expected to Perform in My Graduate Classes?

When I give a graduate course that prepares students for a qualifying exam—in complex analysis or real analysis let's say—then I give regular homework assignments and regular exams. I often grade these myself. On the homework, I expect each student to attempt *every problem* and to make a substantial effort on each one. If a student is having trouble with a problem, then I expect the student to come talk to me about it (or at least discuss it with fellow graduate students). I take a very dim view of a student who hands in only seven of ten assigned problems. I do not mean to imply here that you should just hand in any old thing—even an incorrect solution—just to please Professor Krantz. What you hand in should be *bona fide* mathematics; in particular, it should be correct. You should put forth whatever effort, and invest whatever time, is necessary to be able to achieve the level of excellence that I am describing here.

On the exams in a qual course, I expect each student to attempt most of the problems and I expect a serious effort. You will not pass a qual course exam with a little bit of partial credit here and a little bit there.

A graduate student is expected to expend a considerable amount of effort on the qualifying exam courses and to attempt all of the work in a very serious way. The student should consult the professor and other graduate students regularly and in great detail regarding coursework. Part of what the student is learning is to communicate mathematics, both orally and in writing. S/he is also learning to deal with difficult and daunting work, to organize time, to attack difficult problems, and to make progress in the face of adversity. The student is learning to recognize correct mathematics from incorrect mathematics and to deal with a mathematical problem in the way that a professional mathematician would do.

It is worth noting that graduate courses are usually graded differently from undergraduate courses. The range of grades in a graduate course (and also on the homework and exams) is "A-B-C". Here "A" is excellent, "B" is good/very good, and "C" is failing. This paradigm is nearly universal and you need to understand it. Your program may require you to maintain a "B" average.

In short, graduate school is hard work—and lots of it. Forty years ago, it was commonly said (to undergraduates) that a student should spend five hours studying outside class for each hour in class. Over the years, this dictum has unfortunately dissipated. Over time, it became "three hours outside for each hour inside" and then "one hour outside for each hour inside" and now it's "well, at least do some studying outside of class," but the five-hour rule still applies to graduate work. If you are taking 9 hours of courses, then you should be spending 9 hours per week in class, 45 hours per week studying, and 10 to 15 hours per week on your TA duties. That's a total of (up to) 69 hours per week. That leaves an adequate amount of time for eating, sleeping, and a little relaxation.

A few years ago, a young fellow contacted our Graduate Director to say that he was interested in entering our graduate program in mathematics. He was already an M.D. (i.e., a physician) and he wanted to get an advanced degree in differential geometry so that he could study protein folding. He didn't want any financial aid because he wasn't interested in teaching and he said he could make more than enough money working weekends in an Emergency Room. He seemed like a good bet so we admitted him to the program.

This student was in my undergraduate real analysis class; he seemed reasonably good. As his first semester progressed, things seemed to be going OK, but at the end of the semester he quit. Just like that. He told the Graduate Director that he never realized that the study of mathematics was such hard work.

Bear in mind that the student under discussion already had a medical degree. He was certainly familiar with the concept of "hard work", but the study of mathematics was an entire order of magnitude beyond what he knew. So let me not mince words. If you are thinking of going into mathematics, or already are in a mathematics graduate program, then you are in for long, grueling hours of study. It can be fun and it is rewarding. Mathematics is a great life, but you have to have the personality for it.

3.3. The Qualifying Exams

Our modern concept of graduate school, the Ph.D., and the processes pertaining thereto derives from the German academic tradition of the nineteenth century. In particular, qualifying exams go back at least to that time. The quals are your first big hurdle toward getting the Ph.D. A qualifying exam is usually (not always) a written, timed exam—much like the dozens of other exams you have taken in your life, but it also has special characteristics that we need to delineate here.

Tests in an upper-division undergraduate math class typically check that you know what the important theorems say. Perhaps you are asked to sketch the key steps of the proofs. Perhaps you are asked to perform some straightforward applications.

Not so with the quals. It is not that the quals will be asking you to perform Herculean tasks. Rather, they want to know whether you can think like a mathematician. Consider these dialectics:

- An undergraduate exam might ask you to state the Weierstrass approximation theorem and sketch its proof.
- A qualifying exam might ask you to verify that if f is a continuously differentiable function on an open set U in a C^∞ manifold M and if f vanishes at a point $P \in U$ then it can be approximated in the C^1 topology on a compact $K \subseteq U$ (K containing P) by a function $\phi \in C^\infty(U)$ that vanishes at P.

The second question sounds very fancy, but it is easy if you really understand what Weierstrass's theorem says and if you understand what a manifold is.

- An undergraduate exam might ask you to state Sylow's theorems and sketch the proofs.
- A qualifying exam might ask you to construct a field extension of the rational numbers $\mathbb{Q}$ whose Galois group has subgroups of orders 6 and 15.

Again, the idea is that the qualifying exam wants to know whether you have internalized the ideas and whether you can use them.

- An undergraduate exam might ask you to define Gaussian curvature and compute the curvature of a sphere.
- A qualifying exam might ask you to prove that if Ω is a bounded domain in $\mathbb{R}^N$ with C^2 boundary then there is a number $r > 0$ such that each point $P \in \partial\Omega$ has an internally tangent ball of radius r.

So how do you study for an exam that contains such questions? Typically, you are not going to find questions like these in books. And, even if you do, you are not going to find the answers laid out for you. Let me put it this way: When you study for a qual, it is not enough to learn just the statements of the theorems and the proofs. Of course you *must* do at least that much. And at some schools this basic effort may be sufficient for a low pass. But the new thing that you must learn to do in graduate school is to ask yourself questions. Turn the ideas over in your mind and ask, "Why is the theorem stated this way? Why is this hypothesis really needed? What happens if we change the conclusion from this to that? What would be a counterexample? Why does the proof go like this? It seems that a much easier method would be ..."

The best way to perform this exercise, at least at the beginning, is to study with others. Get together once a week (or more, if appropriate) and try to stump each other with questions just like these. Just formulating the questions is terrific training. Learning to answer them is an even better regimen. And you can work out the answers together. Learning to talk about mathematics is an essential part of any graduate education.

Your department can provide you with copies of old qualifying exams, and it is certainly good practice to do the problems on those exams. The delightful book [SIS] comprises hundreds of problems from the U. C. Berkeley qualifying exams. It is a great resource for students preparing for the quals and a good source of problems in general, but don't delude yourself into thinking that doing practice problems is all there is to studying for the quals. You *must* perform the mental calisthenics described in the last few paragraphs to bring yourself up to speed for the quals. You *must* learn to think about mathematics critically and creatively.

In a later part of the book I will discuss what you must know for a real analysis qual, what you must know for an algebra qual, and so forth. The present section is focused on general principles. The main point is that you must learn to study in a new way. Now *you* are responsible for the material in a deeper and more personal manner. The qualifying exam is posed in order to determine whether you are *qualified to do mathematics at the research level*. Are you the sort of person—do you have the drive, the knowledge, and the determination—to actually *create* new mathematics? This entails critical thinking skills, creativity, a bit of daring, and a willingness to assume risks. You take risks by asking bold questions, by challenging the firmament, and by trying to restructure the subject in your head.

3.4. When Should I Take the Quals?

Most graduate programs have a pretty firm rubric for when you should take the quals (see also Section 3.1). As we have said earlier, it is typical (but not at all universal) for a graduate program to want you to get through the quals by the end of your second year. Of course there will be exceptions, and a good and well-designed graduate program will tailor the curriculum to your background and your needs.

If your undergraduate education has some gaps in it—perhaps you never had a serious course in real analysis—then you will probably have to take one or more basic (undergraduate) courses in graduate school. This will put you a bit off the rhythm and you may have to take an extra semester or so to prepare for the quals. It is easy to imagine other variations on this theme.

Of course it is also possible, even after all the good advice you have been getting from this book and from your mentor(s), that you will fail one or more quals. This is a disappointment, but not the end of the line. Phillip Griffiths, currently the Director of the Institute for Advanced Study in Princeton, used to be a Professor at Princeton University (when I was a graduate student there). He was in the habit of stopping first-year graduate students in the hall, introducing himself, and encouraging us to take the quals as soon as possible. "Heck," he would say, "I flunked the quals the first time I took them. It was no big deal."

Well, you don't want to flunk too many quals and you don't want to flunk them too often. Every program will give you at least two tries, and different people mature (intellectually) at different rates. The quals are not like the thesis.[3] Qualifying exams are just a basic learning situation, one at which you have excelled all of your life. If you apply yourself and follow the advice given here (and *of course* follow the advice of your advisor), then you will certainly get through the quals.

When I was a graduate student, we were expected (if possible) to get through the quals at the end of our *first* year. [Everything at Princeton was accelerated in those days.] I studied like a devil for that year, but when the moment of truth came that spring, I got an attack of the jitters. I couldn't muster the courage to sign up for the quals. I let the deadline pass. With a *huge* sigh of relief, I told myself that I would take the quals the following fall. A few days later, my "Part Zero Advisor" Fred Almgren passed me in the hall and said, "So, Steve, I guess you are signed up for the quals?" With a hangdog expression, I told him that I had not signed up. I wasn't ready. He said, "That's silly. You should take the quals. You'll do fine." I was ready for this. I said, "Well, it's too late. The deadline has passed." Almgren fixed me with a stern glare and said, "I have tenure in this department—and a fair amount of influence. I'll sign you up!" He strode right down the hall and put my name on the dotted line. Fred was a wonderful mentor, and he always seemed to know what was best for me. In this case he was dead right. I *was* ready for the quals, I did well, and I got that chapter behind me.

3.5. How Am I Expected to Perform on the Qualifying Exams?

As I've stated previously, a qualifying exam is not like a calculus test. You will not pass a qualifying exam on partial credit alone. The examiners are trying to determine whether you are *qualified to do thesis work*. Can you see to the heart of a problem? Can you write a proof well? Can you recognize correct mathematics and incorrect mathematics? Can you think critically?

As a result, what the examiners want to see on your qual is a substantial number of questions answered substantially correctly. If there are ten

---

[3]It is actually possible to be as smart as a whip and not have what it takes to do original intellectual work and write a thesis.

problems on the qual, then you had better get seven or eight of them almost entirely correct. If some of the questions ask you to state theorems or definitions, then you had better get them letter perfect—with proper English and all the quantifiers in the right places. What you are learning is a *discipline* and your work had better manifest that discipline.

Qualifying exams are graded according to a strict standard. The exams are formulated to determine whether the student is prepared for thesis work and prepared to go on to life as a professional mathematician. If the Graduate Committee passes a marginal student on the quals, then it is passing a problematic case on to some poor thesis advisor, who will then suffer a long and protracted period trying to cultivate a substandard student. It is also giving reinforcement to a student who probably should not be encouraged to go on. It is not doing anyone any favors to be excessively lenient. Qualifying exams should be fair and reasonable and they should ask questions on basic material that is expected. A qualifying exam should contain no surprises and it should correlate closely with what was taught in the cognate qualifying exam course, but the quals are also a tool for discriminating who should be in the Ph.D. program and who should not, so they should be reasonably tough and thorough.

3.6. What Will My Fellow Graduate Students Be Like?

When I went off to graduate school, I thought I was hell on wheels. I had been a superb undergraduate student, had taken seven graduate courses (while still an undergraduate!), and I was sure that I was the smartest guy around. No?

Well, most of my fellow graduate students at Princeton had pedigrees *at least* as strong as mine, and theirs were from Harvard, MIT, and the University of Chicago (while my *alma mater* was the University of California at Santa Cruz). Suffice it to say that I soon felt like a complete idiot. My colleagues in graduate school not only had the same technical background that I had, but they also had considerable sophistication and exposure to world-class mathematics. Many of them could run circles around me.

Your experience may not be as extreme or as disconcerting as mine, but it is liable to be similar. It is quite likely that you were one of the best, if not *the best*, math student at your undergraduate institution. Such will probably *not* be the case in your graduate program. *All* of the students in a good graduate program were the best, or one of the best, at their undergraduate schools. This is good for you. It will put an edge on your studies. It will set a standard, it will keep you sharp, and it gives you something to strive for (i.e., to keep up with your peers).

My own experience in graduate school was, at least at the beginning, that I learned as much from my fellow students as I did from the faculty. The beginning students were a source of support and empathy. The continuing students offered knowledge and experience. The fact that my fellow students were so smart, and dynamic, and aggressive, made this possible. We all learned a lot from each other.

If you are an American student, educated in the USA, then you may find that the international students in your graduate program appear to be way ahead of you. First of all, students in Italy, France, and other European countries often specialize much earlier than we do in the United States. They have had many more advanced math courses than you will have had. [I was once told that a typical Italian math student graduating from college will have taken 22 year-long math courses. I have not had 22 year-long math courses in my entire life.] Secondly, they often come to our Ph.D. programs having already been through a Master's program in their own countries, and that Master's program is often quite sophisticated; it may entail writing a thesis that involves original work of publishable quality.[4]

It would be easy to talk yourself into the pits if you sit around comparing yourself to the international students. Best is not to compare yourself to anyone at all. You were admitted by the faculty to the program you are now in, so you are definitely qualified for this program; the powers that be have every expectation that you will succeed. You will get through the quals, you will find a thesis advisor, you will write that thesis, and you will get a Ph.D. Your main enemies (should you choose to characterize them as such) are the qualifying exams (which really are not so onerous) and the great unknown (which you must tackle when you write your thesis). Your enemy is *not* your fellow graduate students. In fact they are your friends, and you can expect them to help you just as much as you are willing to help them.

3.7. Am I Supposed to Work All of the Time?

Definitely not. Fred Almgren, my friend and faculty mentor in graduate school, liked to say that graduate students should work four hours per day. What they did beyond that was their own business.

Now one should bear in mind here that Fred had extraordinary powers of concentration. Four hours of work for Fred was like ten hours of work for

---

[4]Of course being an international student at an American university can have its own pitfalls. I have seen international students come here with a grossly exaggerated notion of their own achievements. In some instances I have (with amazement and horror) seen an international student come to a top American graduate program and think that his job was merely to convince the faculty that his Master's thesis could serve as a Ph.D. thesis. I only endeavor to warn people to be suitably humble. When you arrive at your new graduate program, be quiet and listen to what is expected of you.

anyone else, but his point is well taken. You can't do serious mathematics effectively for too many hours in any one day. Maybe eight or ten hours per day is your magic number. But after that you are just kidding yourself; you are just wasting time. You can sit and stare at the book and scribble on a tablet, but you won't accomplish much. And you will also tire yourself out, making yourself less effective for work the next day.[5]

Of course, after you put in your designated number of hours of intense, private work and decide to call it a day, you can still do things that will benefit you mathematically. Go to a colloquium (see Section 3.8) or a seminar (Section 4.1). Hang out in the coffee room and talk about mathematics. Go for a walk with a friend and talk your way through some new theorem that you've learned. Kick around a homework problem with some people. Just being part of a stimulating mathematical atmosphere is good for you.

No matter what, you should certainly set aside time to go to a movie, have dinner with friends, or play raquetball—whatever helps you relax and recharge your batteries. If you try to work all of the time, then you are more likely to have a nervous breakdown than anything else. Don't do it.

3.8. Should I Attend the Colloquium?

Most mathematics departments—certainly all research mathematics departments—have a colloquium. The colloquium is a weekly departmental event, often held late Thursday afternoon, at which everyone (faculty and graduate students) gathers. Traditionally there is a "tea" beforehand followed by a formal mathematics lecture by a mathematician or mathematical scientist from some other department or university.

The colloquium is an essential part of modern mathematical life. Theoretically, *everyone* should attend the colloquium. It is your opportunity to meet new people (either at the tea, through the discussions during the talk, or at the honorary dinner held after the talk) and to get exposure to what is going on in modern mathematics. It is also a ceremonial event at which the entire department gathers and interacts.

So of course you, as a graduate student who is relatively wet behind the ears, should attend as many colloquia as you can, but let me give you a warning. You must not expect to understand every word of any given colloquium. I am a well-read and broadly experienced mathematician who has been in the business for many years, and there is hardly a colloquium of which I can say that I understand the whole thing. If the colloquium is well-constructed and the speaker adroit, then the first twenty minutes will

---

[5]One of the great things we learned as graduate students at Princeton was to *appear* to be goofing off all of the time when in fact we were (secretly) working. To this day, students ask me how I get so much work done when I always seem to be loafing around.

be for everyone, the next twenty minutes will be moderately sophisticated, and the last twenty minutes will be for the experts only. So, realistically speaking, you can expect to understand the first fifteen or twenty minutes of a colloquium—at least to get a sense of what this mathematician is working on and what questions s/he is considering. You should let the rest wash over you; you will get exposure to some new mathematical argot, some notation and some ways of formulating ideas, and what some unfamiliar theorems smell like.

Think about how you learn to play the piano: Of course you begin at the beginning, with hand positions and the most elementary finger exercises. Over a long, protracted period of time you build up to simple pieces, then pieces for which you use two hands, and then finally some serious pieces from the classical piano repertoire. But, even from the outset, you will listen to Vladimir Horowitz and Arthur Rubinstein just to see how it is done.[6] You listen to the masters to gain inspiration and so that you will know what you are striving for. Everyone needs a role model.

Just so, you go to colloquia to see how a working mathematician operates, to get a sense of what is going on in modern mathematics, and to observe how a real mathematical scholar walks and talks. It is also fun, and enlightening, to listen to the questions that are asked both during the colloquium talk and at the very end. You will have here an opportunity to see your teachers in a new light; for now they are students, sitting at the feet of an unfamiliar master.

The colloquium series is an essential part of the life of a mathematics department. You will really be missing out if you neglect to take advantage of this panorama of the mathematics of today. Go with some of your friends, and then (if you are too timid to attend the formal colloquium dinner) go out for pizza together afterward and talk about what you heard. You will gain a new type of lesson about the mathematical life.

3.9. The Foreign Language Requirement

Most Ph.D. programs require you to demonstrate some basic proficiency with foreign languages. The premise here is that an important part of the mathematical literature is not in English, and you should be able to access it. Furthermore, familiarity with languages is part of being a scholar. And it helps, when you attend conferences, to know a little of the argot of your fellow participants.

Today the languages that are most important for mathematics, besides English, are French, German, and Russian. Most Ph.D. programs will ask

---

[6]And, believe me, you will not understand everything they are doing, either.

you to take an exam in one or two of these. The point is usually to demonstrate reading ability and, generally, you are asked to translate a page from a math book (often you are allowed to use a dictionary). The sort of expertise that we are discussing here is, frankly, rather minimal. It is probably not necessary for you to go off and take a language course in order to be able to handle your language requirement. What I and most of my classmates did in order to learn mathematical French was to pull a French math book from the shelf and read it (with the aid of a dictionary).[7] As I have said in so many other junctures in this book, talk to others—both faculty and students—to find out what the language requirement is at your school and what you need to do to prepare for it and satisfy it.

At Princeton and some other elite universities, the foreign language exams are administered orally. In any given year, a certain professor is designated to be the French examiner, another the German examiner, and a third the Russian examiner. If you want to be tested, you knock on the professor's door. The professor pulls a book from the shelf, opens to a random page, hands it to you, and you are expected to translate on the spot. This is what I had to do when I was a student. According to legend, Andrew Campbell, who was a graduate student at Princeton, made the mistake of telling the German examiner (Bochner, a native German) that he had lived eight years in Germany. Bochner said, "OK, your exam is to give me eight meanings for the word 'um'—one for each year that you lived in Deutschland." The student passed, but not without a struggle. He said later that he wished he had lived fewer years in Germany.

Thirty years ago it was fairly common for math graduate programs to allow Japanese or even the computer language `Fortran` as a foreign language. These choices were eventually dropped at most schools. French, German, and Russian are the remaining stalwarts for foreign languages. The hard truth is that, in the past few decades, English has become by far the dominant language in the sciences. It may happen that some day there will be no more foreign language requirement.[8] It is already a bit of an anachronism and it might be hard to defend the requirement in a court of law. It is difficult to maintain that it has a great deal to do with future success as a mathematician (and I even know of cases where an otherwise talented indi-

---

[7]This process usually worked but had its pitfalls. A friend of mine was reading a French math book that began every proof and every discussion with the phrase "Nous sommes ...". He also had the world's worst French/English dictionary. He came to me one day and queried, "Why does every proof in this book begin with the phrase 'We beasts of burden ...'?"

[8]Some math programs have already eliminated their foreign language requirements. Brown has changed from requiring two languages to only requiring one. The University of Texas is in the process of rethinking its foreign language requirement.

3.10. Should I Join a Professional Society?

vidual failed to get his Ph.D. because of trouble with the language exams[9]). But for now it's something with which you must cope.

Some math graduate programs will waive the foreign language requirement for you if you have had a substantial number of language courses in college. Others will give you a pass if you are a native speaker of French or German or Russian. Be sure to inquire about these exceptions if they may apply to you—they could make your life a bit simpler.

3.10. Should I Join a Professional Society?

Absolutely. It is our professional societies that cultivate and nurture the mathematics enterprise. They provide many resources, both for the graduate student and for the working mathematician. There are many professional societies for mathematicians, including the American Mathematical Society (AMS), the Association for Women in Mathematics (AWM), the Society for Industrial and Applied Mathematics (SIAM), and the Mathematical Association of America (MAA). For those with interests in statistics, there is the American Statistical Association (ASA) and the Institute of Mathematical Statistics (IMS).

These organizations hold both national and local meetings. They organize many useful conferences. They provide job information and information about the profession. They publish journals which are staples of the mathematical life (including the *Notices of the American Mathematical Society*, the *Bulletin of the American Mathematical Society*, the *SIAM News*, the *SIAM Review*, and the American Mathematical *Monthly*—there are many more). They publish many books, keep them in print for a long time, and price them attractively. The AMS produces `MathSciNet`—the digital-age version of *Math Reviews*—which is the powerful online archival tool for keeping track of mathematical papers and their authors.

The American Mathematical Society, like the other mainline scientific societies, keeps the government informed of the importance of mathematics and apprised of current mathematical research and activities. Certainly we owe the robust character of the NSF program to support mathematical research in large part to the efforts of the AMS.

If you want to be integrated into mathematical life, be in touch with your colleagues around the country, and know what is going on in the profession,

---

[9]In one notable case, the individual proved some good theorems and qualified for the Ph.D. in every way except that he didn't take the German exam (and the man was fluent in German!). So he went out into the world without the old sheepskin, and he managed to get a job in a good math department. He was a fine mathematician and teacher, but when his chicanery was uncovered he was dismissed. He went to another university, of distinctly lesser quality, and eventually became the Chancellor there, but he still never obtained his Ph.D.

then you should belong to at least one of the professional societies. Students may join at a considerable discount and departments with institutional memberships can provide free membership to students in the department.[10]

3.11. My Duties as a Teaching Assistant

The duties assigned to each graduate student (the quid for the graduate student stipend) vary a great deal. Some graduate students—those perceived to be the best or the most promising—are given a free ride. This means that they have a fellowship and that they have no duties. These days, federal programs provide many fellowships to quite a number of mathematics departments; students supported in these programs have no duties. These fellowships are targeted to American students. A number of international students also receive stipends from their home countries and they are not obligated to teach or do any work either.

Despite the existence of fellowship opportunities, most graduate students must serve as Teaching Assistants (TAs).[11] The Teaching Assistant usually teaches at some (elementary) level. Most often, the TA works for a professor and teaches some of the recitation sections associated with the more formal lectures that the professor is giving. A recitation section is a smaller group (about 25) of students who meet with the TA once (or perhaps twice) per week to discuss homework problems. Truthfully, it is in some ways more difficult to run a recitation section than it is to give a lecture. A lecture is a formal presentation; the person in front of the room has a prepared script and is in complete control. If the professor so chooses, s/he does not even have to entertain questions. However, in a recitation section, the TA is there to answer questions; while many of these will be rote, even humdrum, there will be some that are completely off the wall—completely unpredictable.

I have covered the chapter and verse of how to teach, and in particular how to teach recitation sections, in the book [KRA1]. I shall not repeat those insights here. Let me just conclude by noting that you must take your teaching duties seriously. Learn your students' names. Show them that you care. Make yourself available outside of class. Be fair and even-handed. Your department depends on you to do a good job, the math department's reputation around campus hinges on you and the other math teachers, and

---

[10]In fact the AMS will provide free institutional memberships for *all* graduate students in the member department. The MAA does so for a limited number of students.

[11]Some departments have grading duties for graduate students. These could be meted out *instead of TA duties*—especially if you are an international student and your English needs some work—but grading duties could also be assigned to a TA. Grading duties are pretty easy and you can do the work at your own convenience.

you will be glad to have the respect and admiration from your colleagues that comes from being a good teacher.[12]

3.12. How Will I Be Trained for My TA Duties?

In the old days your teacher training consisted of this: they put a book and a syllabus in your hand and told you to go do your teaching. That's the sum total of the training I received. This was somehow a reflection of the significance that was attached to teaching thirty years ago.

Times have now changed—arguably for the better. Now the math department—indeed every department—is accountable to the Dean for the quality of teaching that goes on. This is true both for faculty teaching and graduate student teaching.

As a result, you will certainly receive at least some instruction about how the math department wants you to conduct your classes and/or recitation sections. At my own university we actually offer a three-credit, semester-long course on how to teach. This course is taught by a senior member of the department, often by the Chairperson. It is required that every graduate student take the course before the prospective TA is allowed to do any teaching. Perhaps other schools will not invest so much time and effort in the teacher-training process. Perhaps they will have a one- or two-week workshop immediately preceding the fall semester of the first year of graduate study. At that time the students are inculcated with basic principles of good teaching.

As I have said elsewhere in this book, you should take your teaching duties seriously. The way that you teach reflects on you, on your department, and on your colleagues. If you need guidance or have questions, don't hesitate to call on the Graduate Director or the person who ran the teacher-training workshop or teaching class that you were required to take. They, too, have a vested interest in seeing you do well.

Many universities have a Teaching Center that offers on-the-spot, custom-tailored help to those struggling to learn to teach. Certainly take advantage of these resources if they are available. Our own Teaching Center films novice instructors and the Director of the Center sits down with each individual to help interpret what is in the film. It is a marvelous (albeit

---

[12]It is all too easy for a bright graduate student, especially a foreign graduate student, to say, "Oh, these American freshmen. They are lazy and ill-schooled and stupid. Definitely not worth my time. I'm going to blow off this whole TA thing." This is a big mistake. Every department justifies its existence through its teaching. If you don't hold up your end of the deal, then you are letting down the whole department. You are giving the Graduate Chairperson and the Departmental Chairperson headaches, and they will remember you for it. A word to the wise should be sufficient.

sometimes humiliating) education, and the process definitely leads to better, and better informed, teaching.

3.13. Organized Labor Among Graduate Students

In the past thirty years there has been unrest among the graduate students at large state universities concerning the TA duties, the level of compensation, the fees they must pay, and related issues. In many cases the crux of the problem seems to have been the right of self-determination. That is to say, the graduate students of course realize that they must perform certain teaching duties and that they are not going to be paid a princely sum for the work, but they want some say in what their circumstances will be: what courses they will teach, at what times, and under what circumstances. Put in other words, the graduate students organized because they felt they had no voice.

There have, from time to time, been some rather dramatic graduate student strikes. The graduate students simply "walked out"—in the sense that they refused to perform their TA duties. Since, at some large universities, a substantial portion of the teaching is performed by TAs, you can imagine that this really put a crimp in the university's style. Hasty negotiations were implemented and the graduate students returned to their posts.

The upshot of these activities is that the graduate students at many universities are now unionized. They now have officers and representatives who negotiate with the departments and the administration on their behalf. Until recently, graduates student unions have been spawned exclusively at public universities, but a few years ago the graduate students at New York University (an entirely private institution) unionized, so unions are now a fact of life at many schools.[13]

Usually membership in the union is voluntary. If you join, you must pay dues (which are usually modest). It tends to be the case that *everyone* benefits from the union's negotiating power—even those who do not join up.

My own view is that unions at a university are a mixed bag. By its very nature, a research university is a meritocracy. Its purpose is to encourage and to reward excellent scholarship. Those who stand out are rewarded more than those who do not, but I must confess that the mass of those who do not stand out tend not to have a voice, and their stature in the institution and

---

[13]It has been the case for even longer that the faculty at many universities are unionized. Again, the unionization is mostly seen at public universities. The faculty at unionized institutions tend to have terrific health coverage and other benefits. The salaries at such institutions tend to be good but not high, the main function of the union being to raise up those at the lower end. Since the present book concentrates on graduate student concerns, I will say no more about faculty unions here.

their quality of life tends to deteriorate over time. The purpose of organized labor is to give such people a role in shaping their lives by allowing them to band together. A university that *does* have a union must still make room for exceptional cases—for the rewarding of scholarly excellence by individuals. It is a delicate balance.

You will have to decide for yourself whether you want to join the graduate student union. It is likely the case that most of the graduate students will be members. If you want to be a part of what is going on and to have a voice in the quality of graduate student life, then you will probably want to join. As with all such matters, be aware of the consequences of your decision before you make a commitment.

3.14. The Research Assistantship

Some departments have funds for Research Assistantships (RAs). Some professors will have grants that provide for Research Assistantships. An RA is a form of graduate student support in which the student's duties are *not* to teach, but instead to help a professor with his research. This could entail library work, computer programming, calculating, or just meeting together and talking (including participating in seminars). Some professors may just ask the RA to do the clerical work for a seminar. Some professors may simply instruct the student to get to work on his/her thesis.

Clearly an RA is a pleasant and benign form of graduate student support. It entails, for the most part, requiring the graduate student to do the sorts of things s/he would be doing anyway.

3.15. The Departmental Staff

While the faculty are out running around the world, giving lectures and attending conferences, it is the staff—the secretaries and other nonacademic employees of the department—who keep things going. These good folks can be your friends and allies during your time as a student.

Who makes sure that you get your mail, your paycheck, or your teaching assignment? It is the staff. Who can secure you a room when you need one (for a help session or a seminar or a meeting)? Who can run interference for you when you can't find your professor, or you need to get in touch with a student, or you need to find out what administrator to consult over a certain problem? It is the staff.

Be courteous and friendly to the staff. They are the people who hold the department together and are there to help you with your work. If they do something especially nice for you, bring them cookies or flowers. At least say, "Thank you". Let them know that you appreciate their efforts and their support.

Chapter 4

Thesis Work

4.1. How Do I Choose a Thesis Advisor?

This may be the most important professional decision you will ever make. First of all, your thesis advisor will have nearly complete control of your life for two or three or more years while you are a student. Secondly, the way your life goes for the first several years *after* you finish graduate school will depend significantly on your thesis advisor, on the advisor's reputation, and on how well s/he looks after you. Thirdly, you depend on your thesis advisor to select a good thesis problem for you and to make sure that you solve it and write it up properly. If it becomes apparent after a while that you have chosen the wrong thesis advisor, then the matter sometimes can be corrected, but it is tricky to do so. The dilemma is best avoided. The remainder of this section will be devoted to fleshing out these points and helping you to select the right thesis advisor for you.

It is of the essence that you work in a subject area, and on a thesis problem, that you like and can develop some enthusiasm for. Just as an athlete will say, "No pain, no gain.", so a Ph.D. student in mathematics might say, "No passion, no thesis." Just so, you also want a thesis advisor who can light a fire under you and can find a problem that will absorb you. You want a mentor who will inspire you to strive for excellence and achievement. You also want a thesis advisor with whom you can work comfortably and have a congenial relationship. If you decide that logic is your thing, and if there is only one logic professor in your department, then the choice of a thesis advisor is easy and obvious; if there are five logicians on the faculty, then you must make an intelligent selection.

Your relationship with your thesis advisor will be a rather personal one. This is an individual on whom you will rely for all sorts of advice and guidance; in addition, this person is going to get you through the thesis and get you started on your career. If you are thinking of working with Professor X, then ask other students what X is like. Is this person accessible, supportive, understanding, a good source of information (both mathematical and otherwise)? Will this professor be a loyal and nurturing mentor? Will this person get you through the program? Does s/he have the resources (through grants or other funding venues) to send you to conferences and get you networked in the profession? Find out how other graduate students chose *their* thesis advisors. Find out which advisors have a lot of (successful) students and which rarely have any.

The nicest advisor is not always the best one. If your advisor tells you that everything you do is great, then you will not be receiving constructive criticism and you will not grow. On the other hand, if your thesis advisor is acerbic and finds fault with everything you do, then you will become demoralized. You want a mentor who is doing great mathematics and who also has a personality that is compatible with your own. You want someone who will get you moving and get you through the program with the best thesis possible.

When I advise undergraduates applying to graduate school in mathematics, I always tell them the same thing: "Go to the best school you can get into, find the hottest, smartest professor around, sign up to work with that professor, and do anything he or she tells you to do." That is what I did, and it worked like a charm. Of course this is not necessarily the best advice for everyone. Some students do not want to go to an extremely competitive place peopled by future Fields Medalists. Some students would rather have a thesis advisor who is a nice guy than one who is a great mathematician. The advice that I am tendering here is for the "best/ideal case".

After you have passed your quals, you are, in effect, in no-man's land.[1] During the qualifying exam phase, people are assigned to look out for you. During the thesis-writing phase, your thesis advisor looks out for you. In between, you haven't got anyone. [Well, if you are lucky, then the Graduate Director and the Graduate Committee are monitoring your progress, but there is not going to be much touchy-feely contact going on there.] It is up to you to pick a research area and a thesis advisor. There are a couple of standard ways to do this.

---

[1]In fact U. C. Berkeley requires students to find a thesis advisor *before* they are allowed to take the quals. This system was set up to prevent students from being stranded after the quals—having no advisor and no prospects.

4.1. How Do I Choose a Thesis Advisor?

Perhaps the most prevalent scenario is that you approach a professor, perhaps one from whom you took a qual class and whom you rather liked, and ask to do a reading course. Typically a reading course is on a subject area not covered by a standard class. You and the professor will pick some texts and some research papers for you to read, and that is what you do. You meet with the professor once per week, ask questions, and discuss the mathematics. As the physicists would say, this is a "toy" version of working with a thesis advisor. You find out what it is like working one-on-one with this particular faculty member, and the faculty member finds out what it is like working with you. If things go well, then it is very natural for you to ask this person to be your thesis advisor. If the reading course does not come to happy fruition, then you will take other reading courses from other professors and try again.

A second method is to take some advanced, seminar-level classes in areas that interest you. Get involved in the subject matter, meet regularly with the professor, try to prove lemmas, work things out, and become part of the process. This activity can lead naturally to finding a thesis advisor (namely, the professor teaching the course).

Another method for getting to know faculty members at the research level is to participate in seminars. You should certainly start attending some seminars by the end of your second year or the beginning of your third. If you become a regular, then you may be asked to give a talk or two. The natural thing is to give a talk on a recent paper or to share a new result with the group (you are *not* expected to present your own original theorems!). Since you are a tyro, you would probably end up consulting one or more faculty members for help with choosing and reading a suitable paper, and this ends up being like a mini-reading course. After you are familiar with one or more faculty members in the seminar, and they with you, then you might ask one of them to be your thesis advisor.

A fourth method, perhaps less prevalent (but used at my own institution), is by way of some formalized talks that each graduate student is required to give. At Washington University we have a Minor Oral and a Major Oral. The Minor Oral is a one-hour public lecture about two or three papers that you have read. Of course you have a mentor for this talk, a faculty member who teaches you to read the mathematics, fill in the gaps in the papers, correct the errors, master the material, and prepare the talk. Our students usually take about six months to prepare the Minor Oral. Then comes the Major Oral. This is a similar endeavor, but on a much larger scale. For the Major Oral, the student might read an entire monograph and/or some really substantial papers. This could take up to a year. You will have a mentor/advisor for this process as well (perhaps the same person as for the Minor, perhaps not). The student then gives a one-hour public lecture. The Minor and Major Orals are part of the mathematical tradition, and part of the culture, of life in the Washington University Mathematics Department. To make a long story short, it is natural for many a student to ask the Major Oral advisor to be the thesis advisor. Frequently the subject area of the Major Oral leads very naturally to the thesis topic.

One thing that almost certainly will *not* work for obtaining a thesis advisor is to just knock cold on some faculty member's door and say, "Hey, Joe, will you direct my thesis?" Most faculty are fairly busy, and agreeing to direct a thesis is a substantial commitment of time, passion, and effort. If the faculty member does not know you, then s/he is going to be quite skeptical of signing you on. If you are lucky, the faculty member might agree to consult some faculty who *do* know you and, if the reports are positive, will then cautiously agree to try you out. What would work a little better, if you are hell-bent on approaching some faculty member blind, is to read some of that person's papers beforehand. At least then you could knock on the door and say, "I've read some of your papers and they gave me these ideas and/or raised these questions. Could we talk about it?" Typically, the faculty member will be pleased and flattered by these attentions, and this could very well lead to something good.

If it turns out, after a year let's say, that the faculty member serving as your thesis advisor is just not working out—by which I mean your personalities conflict, or s/he cannot seem to find a good problem for you, or your styles are orthogonal to each other, or the advisor is simply not capable of giving good and caring advice—then you have to decide what to do.[2] First of all, you should protect yourself. No matter how wonderful, famous, and hard-working your thesis advisor is, you should not restrict yourself to talking solely to this particular person. Present your ideas in a seminar. Knock on some other professor's door[3] and ask whether you can talk to him/her about your thesis problem. Talk to your fellow graduate students. If problems end up developing between you and your thesis advisor, at least then you will have somewhere to turn. In the best of all circumstances, you will then be able to say to another faculty member, "Things are not working out between Professor X and myself. Could we talk about me moving over to work with you instead?" If the faculty at your university are not in fierce

---

[2]It is also possible that the faculty member could decide that *you* are not working out. See Section 6.2.

[3]Always bear in mind, as you read this book, that I am telling you how the world *ought to be*. It is an unfortunate fact that some thesis advisors are very proprietary about the process. They will actually *forbid* their advisees to talk to others and punish them if they do. In sum, you must live in the circumstances that you have.

and continual competition (and, unfortunately, they sometimes are), then something can often be arranged.

You will work for two or three—or perhaps more!—years with your thesis advisor. In the best of all possible circumstances, your advisor will give you a nice, crisp, clear problem and you will go home and solve it.[4] In fact, some thesis advisors (kindness and discretion prevent me from naming any of these) will just give you (the student) a problem and say, "Come see me when you've solved it." This, unfortunately, is rather unrealistic and leads to many unhappy students and to many students who end up dropping out of the program.

A more typical scenario is for the advisor to say, "Why don't you read this paper of mine and see whether it gives you any ideas?" or "Why don't you read this paper and we'll talk about it?" or "Why don't you proofread the chapters of my new book? It's a good way to learn the subject and there are a lot of good problems posed in there." This is fine, and you should do what you are told. Again, don't be bashful. Talk about your reading with *everyone*—with your fellow students, with other faculty, in seminars, and of course with your advisor. It is definitely *not* a zero-one game. It is not as though you can only speak when you have a theorem to show. You can speak any time you like, for almost any reason; and you should.

It is entirely possible to work with more than one thesis advisor; you should relish the opportunity if it arises. This is not really something that you usually can engineer yourself. What is more likely to happen is that you are talking to Professor A about a certain body of mathematics and at some point A says, "You know, Professor B knows quite a lot about this stuff. I'd like to get B involved in our discussions," and, before you know it, you have two thesis advisors. Only one signs off on your thesis, but the other will be on your thesis committee (Section 4.7) and that faculty member's name will be on the title page of your thesis. One of many benefits of such an arrangement is that you will automatically have a knowledgeable second letter-writer when you are seeking your first job.

Some American students—certainly *not* the majority—choose to attend a particular math program because they want to study with a particular professor. I did that; he died the week I got there. It is worth your while to check that the object of your affections is still alive, still active, not retired, and has not moved.

See Sections 4.2, 4.4, 6.1, and 6.2 for some further thoughts on your relationship with your thesis advisor.

---

[4]This is in fact what I did when I was a student, but it was my fourth try; I could not make any real headway with my first three problems. Fortunately, I had a thesis advisor who could go with the flow and keep feeding me different kinds of problems until we found one that clicked.

4.2. Who Is Eligible to Be My Thesis Advisor?

Some universities—especially public institutions—have a subset of the tenure-track faculty designated as the ones who have the right to direct Ph.D. theses. These are sometimes called the "Graduate Faculty" or the "Research Faculty". [When I worked at Penn State, there was a magic moment when I was promoted to the Graduate Faculty.] In principle, any tenure-track faculty member can direct a Ph.D. thesis. Even a professor who is not a member of the Graduate Faculty can be the *de facto* advisor on a thesis, provided a Graduate Faculty member is willing to sign off on the thesis.

By the same token, visiting faculty and instructors generally do not direct Ph.D. theses. As a graduate student, you may be uncertain as to which of the august denizens of the hallways are regular, tenure-track faculty in the department and which are visitors (for a semester or a year or more). The departmental roster—which is usually updated every semester—will make the matter clear, or you can just ask. As with faculty members who are not Graduate Faculty, arrangements can be made for a visiting faculty member or a Postdoc to play a role in directing a thesis, *provided* that some regular member of the Graduate Faculty will sign off on the thesis. At the school where I earned my degree, this phenomenon in fact occurred with some regularity.

The vast majority of thesis advisors are full professors. No rule exists stating that this must be the case, but that is how it usually works out. What is the reason? First, senior faculty members will have more well-developed research programs and will have the experience and flexibility to guide struggling students through the process of writing their Ph.D. theses. Senior professors will know lots and lots of good problems, more than they will ever have time to work on.[5] Secondly, senior faculty members will have the clout and the network of contacts to get their students good jobs. Thirdly, senior professors will have no trouble convincing a Ph.D. committee of the value of their student's work.

There are a number of instances in which Associate Professors or even Assistant Professors have successfully guided Ph.D. students. My thesis advisor was even directing some theses while he was finishing his own graduate work. Unfortunately, having an Assistant Professor as your thesis advisor can lead to some awkward moments. I know of situations where an Assistant Professor thesis advisor was denied tenure and had to leave the university. This left the student somewhat stranded, but usually a tenured Professor stepped in to help the student out. [I also know of cases where a tenured

---

[5]One argument against Assistant Professors directing theses is that they are struggling to find good problems on which to work so that they can publish and get tenure. Why should they want to give away one of them to you?

full Professor thesis advisor died in the middle of the thesis writing. This, too, was awkward. But another Professor stepped into the breach.] In the best of all possible worlds, a more senior Professor will take the Assistant Professor under his/her wing while that Assistant Professor is learning to direct a Ph.D. thesis, but it doesn't always work out that way, and the Assistant Professor can instead find himself/herself in competition with the senior faculty.

To summarize: at most good universities, any tenure-track faculty member can direct your thesis, but, for reasons of seniority and influence and in the general interest of having things go smoothly, it is usually best to have a senior Professor at least as a co-advisor on your thesis.

And now a coda on what to call your thesis advisor. If her name is "Brunhilda L. Schlumpfmeier" then the best, and the safest, thing to call her is "Professor Schlumpfmeier". Some faculty—definitely in the minority— will be uncomfortable with such formality and will say "Call me Brunie" or "call me Hildie". Most will not. At my own graduate school (Princeton) there was an unwritten rule that, as soon as you proved your first theorem, you could start calling your thesis advisor by his/her first name. After you get your Ph.D. it seems reasonable (since now you are a member of the club) to call your advisor by his/her first name. Many a thesis advisor will make it easy by, at some point, saying, "This is silly. Now we are peers/friends. Please call me . . . ."

I am currently a very senior member of the academic world. I have a number of accomplishments to my name, and perhaps am due some respect and deference. Nonetheless, there are certain individuals—august and elderly members of the field—whom I still address as "Professor X". I figure that they are due this courtesy and respect, and that it is up to *them* to say, "No, no. Please call me . . . ."

4.3. How Do I Find a Thesis Problem?

You will probably not be finding your thesis problem yourself, so stop worrying. When I was wet behind the ears and met my potential thesis advisor for the first time, he sat me down and said, "What do you think you want to work on?" I had just read Fefferman's celebrated paper about the multiplier problem for the ball (see [KRA5]) and had found it inspiring. So I said, "I want to work on Fourier multipliers." My advisor said, "That's crazy. It's too hard. I'll give you some other things to do." And he did.

The thesis problems that he gave me were completely unfamiliar and they were all quite difficult. But they were good, they were the kinds of problems that you could jump in and start working on right away, and they had connections to many other things. In other words, they were *ideal* thesis problems. The point that I want to make is that it is virtually impossible for you to come up with such nice thesis problems on your own. You have neither the erudition nor the experience to have any idea how to come up with the right sort of problem at the right level. You have to trust your advisor.

In some cases, your advisor will just hand you a problem. In other cases, you and your advisor will come upon the problem together as your conversations develop (the latter is the methodology that I generally use with my own students). In other cases, your advisor will give you some things to read and the ensuing discussions will lead to a thesis problem. In yet other cases, your advisor will be like a corporate executive who is developing a huge mathematical machine; such a person makes it a point to hand out problems to everyone—from students to junior faculty even to senior faculty who are a bit further down the food chain. You will just be one of the bunch, and you will be given your very own problem.

When an experienced, senior mathematician gives you a thesis problem, then s/he is doing you a tremendous favor. Your advisor is, in effect, saying, "Here is something worth doing and it is at your level. It is a doable problem and you will get a publishable paper out of it. People are interested in this topic and you will begin to make your reputation by solving this problem. Moreover, working on this problem will lead you to other worthwhile things later on." This is something that very few young mathematicians can do for themselves. You are not well enough read and not sufficiently networked in the mathematical community to know which problems are interesting and which are not. You certainly do not know which problems are tractable and which are not, and you do not know which problems fit your abilities. When things do not work out, you do not know what else to try, or how to adjust the problem to make it more feasible.

It rarely happens that a mathematician at *any level* will say, "I am going to solve this problem" and then proceed to solve it. Andrew Wiles did it with Fermat's Last Theorem and Michael Freedman did it with the four-dimensional Poincaré Conjecture. Lennart Carleson did it with the Lusin Conjecture, but John Milnor discovered multiple differential structures on the 7-sphere because something else he was doing wasn't working out. When Kohn and Nirenberg invented pseudodifferential operators, the calculus that they produced was not quite as general as they wanted, but their calculus had some nice properties and it did the job (later Hörmander produced a more general calculus that turned out to be much nearer to what people had sought). When Hironaka proved his celebrated "resolution of singularities" theorem, he failed to get the characteristic p case, which was the one of greatest interest to algebraic geometers (de Jong got it many years later by solving a substitute but closely related problem). Let me assure you that it is unlikely that your thesis advisor will give you a problem and you will just solve it. Most likely you will have a struggle just understanding what the problem is. You may not know the right tools to begin to attack it. You will learn those tools slowly. Along the way a good thesis advisor will make adjustments to the problem, point you in new directions, or teach you new tricks. What you end up with in your thesis will be something worthwhile, something that you can be proud of, but it will most likely *not* end up being precisely what you set out after.

Just as in my own experience (Sections 4.3 and 5.2), you may find that your first thesis problem (or two) just doesn't work out. It may be harder than your advisor anticipated, or it may just be too hard for *you*, or it could be the wrong problem for you. You will have to swallow your pride, file away your notes, and start on something new. You can always come back to the old thesis problems at another time (in my own case, I solved one of my failed thesis problems several years later, my thesis advisor and his collaborators solved the second, and Alberto Calderón solved the third). The main point is to get a thesis problem in a reasonable amount of time, find a job, and start your life.

4.4. My Relationship with My Thesis Advisor

Now you have signed up to work with Professor X on a Ph.D. thesis. What will this entail? The answer is that it will vary a lot, depending on you and depending on Professor X.

Many thesis advisors will want to set up a regular appointment—say once per week—just to keep track of what you are doing. Generally the understanding is, "I don't necessarily expect you to have anything to say when you come by to see me on Tuesday mornings. I just want you to come by regularly so that I know what you've been up to." So some weeks you might come by and say, "I spent the week studying Halmos's book on Hilbert space," or "I spent the week trying to understand the first Mazur/Wiles paper," or "I spent the week trying to prove this lemma but I am stuck." These are best-case scenarios. In the first two instances, Professor X is likely to say, "That's a good use of your time; now read this." Or s/he might say, "Well, that's OK but it's a bit outdated. Why don't you instead look at that?" The third scenario above is really great; this is what a thesis advisor wants to hear—that you are actually *doing* mathematics. Then the advisor can jump in, make suggestions, and try to steer you in the right direction.

Other Tuesday mornings you might drop by and say, "My girlfriend/boyfriend has been in town and I didn't get much work done in the past week." Or perhaps you'll state, "I've been ill and am just recovering." Either of these is fine too. We are all human and have a variety of forces at play in our lives.

You want to have a natural and comfortable relationship with your thesis advisor so that you can be very candid with him/her—so that you can really tell your advisor what you've been doing or not been doing. It doesn't work to be disingenuous and it will only be wasting everyone's time.

I follow a different regimen with my students. We usually do not have a regular meeting time. I tell them to come by whenever they like and we can talk. This is the way my advisor did it and I guess I am carrying on the tradition. Of course this means that sometimes I have to get after students and drag them into my office to find out whether they are actually doing anything, but it seems to work OK. The advantage of this method is that students are not forced to come and stand before me if they have nothing to offer. The disadvantage is that I easily can lose track of what a student is up to, or whether a particular student is actually making any progress.

I was impressed a few years ago when I was reading some obituary articles about Allen Shields (see [DUR], [HAG], [SHA], [ZDDR]). He had a tremendous number of Ph.D. students—at least 28—and was quite an influential and prolific mathematician. One wonders how he could come up with so many good ideas and so many thesis problems for so many students. The answer seems to be that he didn't. Every student had to find his/her own problem. Shields would listen carefully to anything the student had to say and endeavor to respond constructively and helpfully, but he let the student chart his/her own course. That is another way to tackle the problem of directing a thesis.

Your relationship with your thesis advisor really is a personal one. It is up to you, as much as it is up to the advisor, to make it a relationship that works for both of you. The only concrete advice I can offer is this: If you are going to go talk to your thesis advisor, sit down for half an hour beforehand and think about what you want to say. It does no good, and makes a poor impression, for you to go into the professor's office and just hem-haw around—never coming to the point. Don't waste the professor's time. Use it constructively. S/he will appreciate your courtesy and you will feel better about yourself as well.

If you are stuck on what you are doing, or if you are discouraged by what you are doing, or feel that you are not making any progress, then *please say so!!*[6] Your professor is not a mindreader. If your advisor is going to help

---

[6]I had a good friend in graduate school who was getting more and more depressed about his work. He was making no progress and his problem was making less and less sense to him. One week he went in to see his thesis advisor and let it all hang out. He told the advisor *everything*—how distraught he was, how little progress he was making, the troubles he was having with his wife. He described the whole enchilada. The thesis advisor sat there quietly, chain smoking and

you, then you must say directly where you are, where you are coming from, and where you are going.

I am going to conclude this section with an uncomfortable *coda*. Even when I was a child, my parents (neither of whom had a college education) warned me about professors who might appropriate my brilliant ideas and go off and publish them—without giving me any credit. On the face of it, this may sound ridiculous, but in fact it *can* happen. A student will do something really brilliant—something the professor wishes that s/he had done—and the advisor convinces himself/herself that this never would have happened without his/her good offices, guidance, and extensive knowledge of the literature. There may even be some truth to these ostensibly self-serving feelings. But you can imagine that some uncomfortable and even nasty confrontations can result. I know of one incident where the professor basically said to the student, "This result is too good for you. And you never would have stumbled upon it without my guidance. I will publish it under *my name alone*,[7] and you can do something else in your thesis."

Well, this is really a shame, and it puts the student in an extremely difficult position. Let me assure you that this almost *never* happens, but it again illustrates the value of sharing your ideas with other graduate students and *especially* with other faculty. It is always the case—at least in the academic world—that the best protection for your ideas is the academic *community* itself.

4.5. How Do I Work on My Thesis Problem?

This is *neither* a trivial nor a naive question. The answer is not, "Just sit down and do it." I had a friend in college who did all his homework assignments, and his takehome exams as well, by poring through every book in the library until he found something that looked like a solution to the exercise on which he was working. This worked fairly well. He received reasonably decent grades but he did not really learn much of anything, because he never put his mettle to the test. You develop your brain by banging it against things, by stretching it, and by challenging it. This means that you must do the work *yourself*.

With the thesis work, this advice is even more decidedly true. You certainly are not going to find your thesis problem, nor anything like it, in a book or even in a journal. You must do it yourself.

---

nodding his head. At the end he said, "That's fine. Come back and see me again next week." My friend got more depressed. One hopes that this experience is not typical.

[7]In this case the student appealed the matter to the AMS Ethics Committee and the professor was disciplined.

The short answer to the question is: You work on your thesis problem by sitting in a quiet place and calculating. You try things and then modify them and jiggle them and then try them again. Fill dozens and dozens of pages with your speculations and trials and scribblings and conjectures, then throw them away and start again. During this process, you are constantly talking to people, going to seminars, writing *e*-mails, and asking questions. You immerse yourself in the problem and swim around in it until you find something that floats—something to latch onto. Gradually, you develop that handhold into a thesis.

Charlie Fefferman, Fields Medalist, once said that a good mathematician throws away 90% of his work. I would put it at 99%. Anything worth doing you are going to generate yourself, and only after many, many hours of trial and error. You start with simple examples—the simplest you can find—and work up to more sophisticated ones. You keep changing the parameters until you can make the thing come out the way you want. And of course you continually talk to people—especially (but not exclusively) to your thesis advisor. You find more and newer and better ways to turn the ideas over in your head, put them back together in new ways, and find new insights. Always remember that your progress will be by small increments; you are not going to solve the thesis problem all at once, in a great flash of insight.

Also bear in mind that the path to a thesis is not a linear one. The police chief in the Pink Panther movies—in moments of extreme frustration—would chant to himself, "Every day, in every way, I'm getting better and better." Let me assure you that, in working on your thesis, you will *not* get better every day. Some days you will seem to learn something new and get closer to the goal. Other days you will have to tear up what you did the day before and try again. There will be days when you seem to be blocked and others when it appears as though you are on a fool's errand. Don't let it get you down. This is the life of a mathematician. Your entire career is going to be like this. At least now you have senior mentors to guide you.

Learn good work habits. When a calculation finally works out, write it up carefully, number the pages, date it, and put your name on it. File it away in an organized manner so that you can find it again. Keep a daily journal. Record in it what you have tried, what works, and what does not. Be aware of the fact that, when you do a calculation or make a discovery, it will seem as plain as day and something you will never forget. Sadly, you will. And you will forget it most surely exactly when you need it. So write it down. Often things that you tried and did not quite work out are just as valuable as things that are beautiful and glow in the dark. Keep a record of everything you try. This will be valuable both for psychological support and also as an archive of your efforts. When he was young, Lars Hörmander

4.5. How Do I Work on My Thesis Problem?

kept a mathematical diary: at the end of each day he would record what he had learned and what he had not. He won a Fields Medal in his early 30's, so perhaps there is something to be said for good work habits. Gauss followed just the same regimen and his efforts are also remembered well.

The more you do mathematics, the more you will treasure concrete examples. I would say that most of my best work is based on just a few examples that I return to over and over again. Always remember that we learn *inductively* (going from the specific to the general) rather than *deductively* (going from the general to the specific). The deductive mode is highly appropriate for *recording* mathematics, but it does not work for *discovering* mathematics. You discover, and create, mathematics by starting small, by doing little calculations, and then working up to more substantial and meaningful calculations. The aggregate of many calculations can become an insight and that insight might turn into a conjecture. Even more effort might transmogrify that conjecture into a theorem. It is a fantastic process and a marvelous intellectual tradition to become a part of. The writing of the thesis is your entrée.

The most important skill that you must learn in order to become a mathematician is to overcome the natural mental inertia with which we are all endowed. The easiest thing in the world would be to sit at your desk and stare forlornly at the thesis problem, constantly chanting the mantra "Poor little me. What am I going to do?" Unfortunately, this technique does not work well to solve a problem. The journey of a thousand miles begins with the first step, and it is up to you to take that step. Your thesis advisor may show you what that step is. S/he may even place your foot on the mark. But you've got to do it.

One of the main things that I must teach my Ph.D. students is that the mathematics that you need in order to do creative work is not written down. You may find references that will point you in the right direction, or give you that boost or inspiration that you need, but you will almost never find just that little lemma, or just that incisive theorem, that you seek by paging through books or journals. You must cook it up yourself. The process requires an entirely new mindset. You must learn to jump right in and try things. Talk to people. When you are stuck, collar a friend, drag him/her to the blackboard, and tell your tale. I can't tell you how many times the process of explaining a problem to someone else has caused me to rethink the situation and, ultimately, to resolve my dilemma. The main point is that now you are generating the mathematics yourself. You will start with very small, often infinitesimal, increments and you will work up to decisive steps that will lead to a thesis.

There is much to relate about the process of learning to do mathematics and yet there is hardly anything that one can actually say that will amount to much more than, "Go forth and sin no more. Sit down and begin. You will learn by doing." Be assured that every mathematician wrestles with this process every day. As you write your thesis, you are learning the moves.

This is an entirely new *gestalt* for you—one that you must learn by painstaking trial and error. However, it will become the *matière* of your professional life. It is your bread and butter, the set of skills that makes you a mathematician. You may as well start developing them now.

4.6. How Do I Write Up My Thesis?

Let's face it: the longest, most protracted mathematical writing you have ever done up to now is the solution of a long homework problem. Right? Maybe you wrote a senior thesis, but that is fairly low-key mathematical writing. Quite episodic. Your Ph.D. thesis will be 75 or more pages of deep, dense, connected, and original mathematical writing. The very structure of such a piece of writing boggles the mind. It is a monumental task, almost entirely unfamiliar to you. You will need *at least* three months of very hard, concentrated work to do it.

Mathematical writing is a very particular kind of writing. It makes special demands on both the writer and the reader. The book [KRA2], and the references therein, can give you guidance on how to be an effective communicator and an effective creator of mathematical writing. You may also get some useful tips from [KRA3] regarding typesetting and graphics.

I always insist that my graduate students leave nothing to the imagination in their theses. Every detail must be there. We professional mathematicians, in the interest of brevity and also as part of the *savoir faire* of our discipline, are wont to say things like "by a slight variant of Theorem 3.7 in the paper of Ignatz" or "using the general approach of Hilbert ...". I simply will not allow this in a Ph.D. thesis. The student must write out all the proofs. Period.

The only way to get the writing process going is to begin. Start small. Generate a broad outline of your thesis. The first pass can be a very vague adumbration, with just the key topics laid out. Then develop the outline by gradually adding detail. Show it to your thesis advisor just to make sure that you are headed in the right direction.

Now create a *very detailed* outline. This draft should actually list every definition, every lemma, every theorem, and every example. Don't write them all out. Just indicate each one with a couple of words: "the covering lemma" or "the spectral sequence argument" or "the completeness axiom".

4.6. How Do I Write Up My Thesis?

This detailed outline might be ten pages or more. In this manner you can get the project completely organized before you start the serious writing. You can make sure that all of the logical steps are there. You can endeavor to plug all of the gaps. Part of this outline is to list all of the references. For now, just indicate each one with a one- or two-word abbreviation. Again, show this effort to your thesis advisor. Listen carefully to whatever criticisms are tendered.

Plan for your thesis to have a very detailed introduction. Part of writing a good thesis is putting it into a context. Give the history; tell who invented the subject and who contributed to it. Discuss what has been tried and what works. It is probably best to write this introduction last, because then you will have the whole business crammed into your head and it will be a pleasure to write a five-or-more page explanation of what it is all about.

Now it is time to actually write. You don't have to write the thesis in strictly logical, linear order (because you will be using a computer and some version of TeX[8]). Just get started *somewhere*. Once you get into the flow of it, you will find that the writing gets easier and easier. You will stop worrying about writer's block and be carried along by the passion of the mathematics. One thing you will find, and let me assure you that every mathematician everywhere must contend with this problem, is that certain lemmas or steps that you thought were obvious will turn out to be not so obvious. You will find gaps and roadblocks and difficulties with your work. Don't panic. This is why you have a thesis advisor.

When I was writing my thesis, I came across a serious gap in one of my arguments. I was devastated. I spent several weeks avoiding my thesis advisor. If I saw him coming down the hall, then I hid in the bathroom. This was idiotic. When I finally summoned my courage and went to tell him of the trouble, he was very kind and very helpful. It turned out that he didn't know how to do it either (he had some notes on the calculation and they turned out to contain an error as well), but he told me how to handle the situation, I wrote up what I had, and everything turned out fine.

Your thesis should be written up in some flavor of TeX. These days there really is no other reasonable choice. TeX, as you may know, is a computer typesetting language for mathematics. It gives you publication-quality output and allows you almost unlimited flexibility in formatting, font choice, and layout.[9] You can learn TeX quickly and easily (in about

---

[8]This could be LaTeX, $\mathcal{AMS}$-TeX, $\mathcal{AMS}$-LaTeX, Plain TeX, or one of the other variants. The books [KRA2], [KRA3], and [SAK] explain all of the typesetting systems' nuances.

[9]Perhaps you are a person who is computer-phobic, or who doesn't have access to a good computer and laser printer, or perhaps simply doesn't have access to TeX. Another possibility is that you just don't want to be bothered to learn TeX. Not to worry. You can certainly hire someone to do your TeXing for you. But be forewarned that this can get rather expensive—your

three hours) using one of the books [SAK], [LAM], [SPI], or [GRA]. There are many other useful references as well, some of them listed in the back of [SAK]. The reference [KRA3] can also give you some guidance in how to use TEX.

Emotions can run high over which version of TEX one ought to use, and I don't want to get into that imbroglio here. Many beginners, especially the gearheads, gravitate to Plain TEX because it gives them the greatest artistic freedom to create a typeset page with that "special flair". That is *not* what you are striving for in your Ph.D. thesis. Let me just say that there are strong arguments for using either LATEX or $\mathcal{A}_\mathcal{M}\mathcal{S}$-LATEX, as these macro packages are more structured, leave the user less latitude, are less prone to error, and (most importantly), can most easily be converted to other software formats. For more information, see the web site http://www.ams.org/jourhtml/latexbenefits.html.

Mathematics is hard; learning TEX is relatively easy. It is all too common for a math graduate student to get seduced by TEX and to spend more and more time learning programming tricks that have no bearing on a mathematics education. Always remember: you will be hired and promoted because of your mathematics, not because of your TEX skills. Using LATEX will tend to stifle any high-tech creative tendencies that may crop up in your ganglia, but it will get the job done and leave you more time for your studies.

A full implementation of TEX from a commercial vendor can be rather pricey. Likely as not, your university will have a site license for some version of TEX and you can obtain a copy for free. Also MiKTEX, a particularly attractive and versatile version of the software for the PC, is available for free on the web.[10]

Many theses will contain graphics. Of course you can render each graphic, by hand if you like, on a separate sheet of paper, or you can draw

---

thesis will probably be at least 75 pages and the charge is liable to be at least $10 per page. My own university used to type Ph.D. theses for free, as a service to the students. That is quite unusual, and we don't do it anymore either, so don't expect such largesse from your school.

[10]MiKTEX comprises Plain TEX, LATEX, and $\mathcal{A}_\mathcal{M}\mathcal{S}$-TEX, as well as some other standard variants. The lovely MiKTEX command texify analyzes the source code in a TEX file, determines which type of TEX it is, and then compiles and produces the output *.dvi file. MiKTEX is also particularly good at handling graphics. You can find out more about MiKTEX, and download it too, from one of the following web sites:

http://members.tripod.com/~upem/miktex.html
http://www.miktex.org

or from the ftp site

ftp://ctan.tug.org/tex-archive/systems/win32/miktex/1.20/index.html.

The reference [KRA3] will give you more information about TEX sources, both for PC and Macintosh computers. The TEX Users Group (TUG) at http://www.tug.org has many resources and implementations of TEX. Finally, the American Mathematical Society (at http://www.ams.org/tex/) has many useful links.

4.6. How Do I Write Up My Thesis? 69

each graphic with a paint program or some other piece of software for rendering art. Then you can interleaf each graphic with the text that you create in TeX.

The best way to proceed, however, is to create each graphic with a drawing program like Corel `DRAW` or Harvard `Graphics` or Adobe `Illustrator` or `xfig` (in `UNIX`). Commercial products, like the first three of these, will save each graphic in a proprietary graphics language, but you can export each graphic to `PostScript`, and then it is a straightforward matter to incorporate the `PostScript` graphic into your TeX document.[11] The book [KRA3] will give you some guidance as to how to do these things. In this way you get a seamless, single document that contains all of your text and all of your graphics in a professionally prepared presentation. You can exhibit each graphic at any magnification, located just where you want it on the page, and you have complete control over the spacing around the graphic. Once you get accustomed to this working environment, you will use it in preference to all others.

And now a warning: Most universities have very strict rules as to how a thesis should be prepared. Go to the Graduate School Office and obtain the little handbook on thesis preparation.[12] It will tell you, in excruciating detail, how to set the margins, what font type and size to use, how to display graphics, how to format the bibliography, and so forth. It will also tell you how the thesis should be printed out (and on what kind of paper), prepared, and bound.[13] Let me assure you that each university employs some closet martinet to check these details assiduously.[14] If your thesis does not conform, then it will be rejected and you will have wasted a great deal of time, expense, and effort.

At the risk of belaboring the obvious, let me stress that a Ph.D. thesis is *not* submitted electronically. You will submit a specified number of hard

---

[11] Note that many computer algebra systems—such as `Mathematica`, `Maple`, and `MatLab`—will do a splendid job of producing graphics (especially graphs of functions and other data displays). You can then output these graphics to `PostScript` for inclusion in a TeX file.

[12] The rules will be different for every university. Be sure to get a copy of *your university's* handbook.

[13] Please note that you will *not* submit your thesis stapled together, spiral bound, or in an ACCO folder. There are very specific requirements for the preparation and binding of a thesis. You will probably have to pay someone to do this for you.

[14] There is actually some logic to this madness. Traditionally, the university itself was the sole archive for Ph.D. theses written under its auspices. The university wants each thesis to endure, and to be accessible and readable, for hundreds of years. That consideration is the provenance of all of the rules. For many years there has been a separate third-party service called UMI in Ann Arbor, Michigan that archives Ph.D. theses on microfilm; it archives some Master's theses as well. Furthermore, it provides individual institutions with electronic copies of their own theses so that they may be made readily available to the university community. The web address is `http://wwwlib.umi.com/dissertations`. Also many theses are posted on the web and elsewhere. But the strict rules for preparing theses endure.

copies, printed and bound in a specified way. These old-fashioned and hidebound rules exist in part for the sake of tradition, and in part because hard copy is the only medium that we really know how to archive.

4.7. My Thesis Committee and My Defense

It is customary for there to be a committee to evaluate your thesis. Your thesis advisor is the chair of the committee. Then there are a couple of other math faculty on the committee and one or two faculty from *outside the mathematics department*. What does this mean?

It is a strange tradition. The purpose of the thesis committee is to prevent certain eccentric faculty from ramrodding weak students through the Ph.D. program. The committee provides a system of checks and balances, and the committee members from outside the department are meant to keep the math department honest.

Of course the outside members do not have much chance of understanding the thesis in any detail, but if they are good scholars, then they will have a sense of whether this is the real stuff or just a bunch of smoke and mirrors. They will be able to tell from your formal presentation (see below) and from the flow of ideas whether this is genuine scholarship of good quality.

Generally speaking, it is up to the Ph.D. candidate (namely, you) to ask members of the math department to serve on his/her committee. This is no big deal: you ask them and they say "yes". It is part of their job. Getting the outside committee members is a bit trickier. At my own university, the Departmental Graduate Office will find the outside members for you. At other universities (such as UCLA, where I used to teach), the department has a standard list of friendly faculty in other departments whom you, yourself can ask.

Theoretically, the members of the committee are supposed to read your thesis. Certainly your advisor will read it, line-by-line, and will offer copious and helpful criticisms. You will in fact give your advisor the draft of your thesis many weeks before the defense (described below). The other committee members are given a week or two to look at the thesis (this matter is strictly regulated—be sure you know the rules when the time comes). The math department members probably will look at parts of the thesis. The others probably won't. Then comes the day of truth.

The "day of truth" is the thesis defense. This is a ceremonial occasion[15] at which you present your results to the public. Here "the public" consists of your committee and those members of the math department, and perhaps

---

[15] In the Spanish Department at UCLA it used to be the case that a big part of the thesis defense was for the candidate to prepare a gourmet Spanish dinner for the examiners.

4.7. My Thesis Committee and My Defense

some friends or classmates or family,[16] who choose to attend. You are given about an hour to present the essence of your ideas.[17]

The audience is then allowed to ask questions. Next, the visitors are asked to leave the room and the candidate is left with just the committee. The committee is then allowed to ask some more pointed questions, and often it does. Fortunately for you, your thesis advisor is right there and (you can bet) is on your side. Your thesis advisor would not let the thesis defense take place unless s/he thought that *both of you* were prepared and your thesis advisor will help with any tough points that may come up. After the committee has grilled you, so to speak, then you (the candidate) are asked to leave the room. The committee then discusses the case *in camera*, often asking questions of the thesis advisor. [To repeat, the committee needs to be convinced that this is a worthy and substantive body of new work—work that will be a credit to the department and to the university. If your thesis advisor is in good standing with his colleagues, then the whole process should be a formality.]

And that is the final step. This is an opportunity for your thesis advisor to sum things up for the committee and point out the merits of the thesis, and of you. Then, assuming that all has gone well, the committee calls the candidate back into the room and the members shake the candidate's hand. Usually the entire process goes quite smoothly and everyone comes away happy.

Despite all of the pomp and ceremony, the thesis defense is serious business. You are, by means of this lecture, showing to the world that you have performed original scholarly work and you are given this opportunity to explain what it is. You must be in control of everything that is in the thesis and you must be able to explain all parts of it—where the problem comes from, why it is interesting, what techniques were used to solve it, how the proof works, and what you plan to work on in the future. Preparation for the thesis defense is a serious and protracted exercise. You want to show both yourself and your advisor proud. Of course your advisor will help you with the process. Consult him/her frequently as the need arises.

I *have* seen situations—not many—where a student fails a thesis defense. This almost never happens, because a responsible thesis advisor will make very sure that the student is ready for the defense. If the student is *not* ready, then the defense will never take place. At least this is the way it should be.

---

[16] A recent Ph.D. defense at my university was jam-packed both with family and with members of the candidate's Lutheran church. There was even a cameraman shooting photos for the Lutheran Church Newsletter. This is most unusual, and you should not plan on such a dog-and-pony show at your own thesis defense.

[17] This is a good time to dress nicely. You don't need to wear a top hat and tails, but wear the sort of clothing you would wear to your maiden aunt's for Thanksgiving dinner.

Those rare occasions in which a graduate student engages in the defense and fails happen because it becomes rapidly clear during the presentation that the student really doesn't know what s/he is talking about. The examiners conclude that the thesis advisor has, in effect, written the thesis for the student and is using the student as a cat's paw to get the results validated. [Or, on even rarer occasions, the committee concludes that the thesis does not really have much in it.] And this is what a thesis defense is for. The student really does not merit a Ph.D. and the plug gets pulled. The good news is that such a student is usually allowed to re-assemble forces and make a new presentation; things usually go fine the second time around. Let me emphasize, once again, that nobody has it in for you. Everyone wants to see you succeed. But you must, of course, meet the expectations of the program.

The tradition for thesis defenses in Sweden is both remarkable and charming. First of all, the candidate is present but does not participate. Instead the faculty enlists a famous foreign professor to read the thesis and to come in to make the presentation on the student's behalf.[18] One Swedish professor is then assigned to ask pointed questions of the foreign presenter. Another Swedish professor is assigned to crack jokes and make fun of the whole proceeding.

It is a sorry fact that the role of the second Swedish professor has been eliminated in recent years, but it is heartwarming to see that the Swedes have had some fun with this otherwise august occasion.

I don't mind telling you that I was a nervous wreck both before and during my thesis defense. It is, however, one of those rites of passage that we all must live through and I look back on it now with fond memories. One of my fellow students gave his thesis defense in one of the seminar rooms in the Fine Hall tower at Princeton. In the middle of his presentation he turned to face the audience and he could see through the window a sign (lowered on strings from the floor above) that said, "Error on page 72 of your thesis. You flunk." He began to laugh, and the examiners turned around to see what the joke was. When they saw the sign, they quickly grabbed for copies of the thesis. It had only 69 pages. I think this incident is very much in the Swedish tradition and I applaud it.

---

[18]Analogously, in the Canadian system it is common for the university to enlist an outside reader from another institution to read and evaluate the thesis.

Part 3

Sticky Wickets

Chapter 5

Practical Difficulties

5.1. Should I Hold an Extra Job While I Am a Graduate Student?

Perhaps you are married. You might have a child. Or you might have a girlfriend or boyfriend who likes to eat at fancy restaurants and drive sports cars. You are tempted to get an extra job waiting tables in a restaurant or washing cars or doing clerical work—just so you can have some extra spending cash. Please resist this temptation if you can.

Being a graduate student is a full-time job. You need to concentrate all of your effort on your studies. While your job of being a TA can be a tedious chore, at least it is mathematics and the duties are performed at the university. It's still part of the academic game.

Working at an outside job is a distraction, it is tiring, and it is mind-numbing. It will not only use up your valuable time but also it will exhaust you and dull your wits. You will be making your job of passing the quals and writing a thesis vastly more difficult if you inject complications like this into your life. It really would be better to get a loan from your parents, or from a bank, than to shoulder the burden of getting a job on the side while you are a graduate student. Sometimes your university will lend you money.

The idea of taking out a loan while you are still in school may be frightening. As my parents put it, "Why would you want to start life with such a burden?" There is another way to look at the matter. First, you should understand that payments on a college loan are deferred; you don't begin to pay it back until after you graduate and have a job. Once you are out in the economic system and have a regular income for a few years, the monthly payments on the loan will seem relatively trivial. You'll have it paid off before you know it. So don't be bashful about exploring the option of getting a loan if economic necessity so dictates. Best, of course, is to structure your life so that you can live within your means.

And now a final warning: Some graduate programs, such as my own, have strict rules about outside jobs; namely, you are not supposed to have one. If you are the holder of one of the many federal graduate fellowships, then the rules are even stricter. You could lose your TA-ship or fellowship by moonlighting on the side. So don't do it.

One exception to the stern admonition of the last paragraph is this. Everyone seems to be willing to look the other way, and all in all it seems to be rather harmless, if you want to do some undergraduate tutoring. Often you can arrange to do the tutoring late at night, when you are probably too tired to think about mathematics anyway. Current rates for this service range from $20 per hour and up. It's a relatively painless way to earn some extra money, and it has the virtue of still being mathematics and still being done in an academic environment.

5.2. What If I Can't Solve My Thesis Problem?

Join the club. Most mathematicians are bipolar disorders waiting to happen. We spend our lives wrestling with problems that we cannot solve. With your thesis problem, you are getting your first taste of this syndrome.

Your thesis problem will not be like a calculus problem—something that you just sit down and do. It is supposed to be something meaty, new, and original, something that might lead to one or more publishable papers. *Of course* you are going to have trouble with it. That is why you have a thesis advisor.

As a professional mathematician, if you are going to survive, you learn to adjust your goals, add hypotheses, change the question, weaken the conclusion—in short, do whatever is necessary to get somewhere so that you can state a theorem and prove it. With the help of your thesis advisor, you will do the same (probably several times) with your thesis problem.

Mathematics is deep, complex, and difficult. Craig Evans of the University of California at Berkeley has told me that 99.9% of what he tries fails—and he is a very successful mathematician by any measure! One of the reasons that mathematicians tend to be eccentric is that we confront failure and disappointment every day, and try to find ways to surmount it. For the sake of success in mathematical research, it is much more important to be able to cope with the frustration inherent in the process than it is to be "quick" or "brilliant". To be sure, the top people are often successful, lucky, quick, insightful, and (overall) just dazzling. But the mathematician who succeeds, generally speaking, is the one who learns to keep plugging away, building up technique and insight as many things fail and a few succeed. One of the joys of doing mathematics is that those things that we do manage to prove are of permanent value. They will be as true 100 years from now as they are at the moment of their discovery. Many of them will become part of the mathematical infrastructure, even if they are not explicitly stated (with your name attached) in books.

The struggles that you have with your thesis, and will continue to have throughout your career, are no source of shame or concern. It is part of the process. You learn by doing, and in this case you are learning to turn your problem into one that you can solve (and that is still worthwhile, and worth writing up, and perhaps worth publishing). This is an art, one that you need to cultivate. Fortunately for you, your thesis advisor is no doubt a past master and can get you through this. You merely have to trust him/her. I will repeat something that I have said elsewhere in this book: You don't need to have a theorem in your hand to go talk to your thesis advisor. Any excuse is a good one. Knock on the professor's door and tell your tale. You are bound to get something good out of it.

5.3. How Much is Enough for a Thesis?

Assuming that your thesis advisor is a member in good standing of the math department, the answer to this question is really up to him/her. Sometimes (not often) a thesis will be centered around some blockbuster theorem. Once you have proved it, you know you've got a thesis. More often, the thesis is an agglomeration of many results. Sometimes the thesis will be about more than one topic—although the topics will likely be closely related. It is really up to your thesis advisor to finally say, "OK, you've got enough. Write it up and that will be your thesis."

I will tell you frankly that there are few theses that contain everything that the thesis advisor originally envisioned. Usually calling a halt to the process is a matter of the thesis advisor finally saying, "You have proved your mettle, you've been here long enough. Let's get you out of here."

The bottom line is this: You, yourself have no way of judging the amount of material sufficient for a Ph.D. thesis. Ultimately, it is your advisor who is putting his/her reputation on the line by placing you before a thesis committee, so it is s/he who decides.

I hate to keep saying this, but if you are constantly talking to and consulting regularly with your thesis advisor, then there should never be any doubt as to how far along you are in your thesis work. I can still remember the day when I stared at a certain integral, performed an integration by parts, manipulated the result around for a while, looked at the upshot and said to myself, "Hot damn. I think that's my thesis!" I could do so because my advisor had made it crystal clear to me what was expected and what my goals were. A friend of mine (one of my undergraduate mentors) told me that once he was able to write down a certain matrix of 0's and 1's—thus creating a specific example of a particular kind of operator—he knew he had his thesis. He also had a great thesis advisor who had made things crystal clear to him. If, at any stage of your thesis work, you don't know where you are—then just ask.

5.4. Why Does a Graduate Student Leave the Program?

When graduate programs are being assessed, a question that is usually asked is, "What is the attrition rate?" That is to say, what percentage of those admitted to the Ph.D. program leave without attaining that august degree? The percentage, sad to say, is fairly high at most schools. An attrition rate of 30% or 40% is not uncommon. At Washington University, in the mid-1980's, our attrition rate was less than 5%, but the circumstances were unusual. First of all, most of our Ph.D. students were hand-picked and identified through personal contacts; they came to us already having Master's degrees from top universities in Italy, Spain, Argentina, Poland, and China. These students hit the ground running, had many of their quals waived, got involved in research almost from the moment they walked in the door, and got through the program both expeditiously and (virtually) without fail. Moreover, in those glorious times, our graduate students had no duties. *They did not have to teach!* They could devote all of their time to learning to be mathematicians and they did so with vigor and effectiveness.

The situation described above almost never obtains today—not even at Washington University. Today most of our students are American; their levels of preparation are more uneven and their levels of motivation cover a wide range. They all have serious teaching duties now and many of them have spouses or significant others; many have children.

There are quite a number of reasons that a student may leave the program, some personal and some mathematical. A certain percentage of our students find that they just are not up to the level and quantity of work required in a mathematics graduate program. They leave at the end of the first year. Another bunch just cannot get through the quals; they leave at the end of the second year. Of those who get through the quals, the vast majority manage to write a thesis and get a degree, but there are those who get a divorce, or have a death in the family, or run out of time, or run out of money, or lose the desire, and have to quit. The university has many mechanisms—including leaves of absence and loans—to give such people a helping hand. These folks tend to be much fewer in number than those who drop out in the first two years, but there are a couple every year.

There are also students who just lose the spark. They decide that mathematics is not for them. They look ahead to three or more years of writing a thesis followed by six years as an Assistant Professor and another six or more as an Associate Professor and say to themselves, "This is not how I want to live my life."

I have counseled students who have run into all of these impasses. I have been able to help a number of them—convince them to stay in the program, get a degree, and have a good career in mathematics, but for some it is clear that I would be doing them no favor by convincing them to stay. It becomes evident, after some discussion, that the student's life is taking a new direction and that a long program of deep and protracted study is not in the cards. For such a person, it is better to see the writing on the wall and to make a change.

5.5. I Don't Seem to Know Anything!

I introduced this theme in the first few lines of the Preface and now I am returning to it. I frequently find myself advising my Ph.D. students with respect to this question. It is a constant torment for serious scholars at every level, so let me make a few things perfectly clear.

Mathematics is a huge and diverse discipline. People like to say that David Hilbert (1862–1943) was the last mathematician to have mastered all parts of mathematics. He certainly made significant contributions to all of the chief areas in the subject. It is safe to say that, today, there is nobody who knows the whole field. Mathematics is just too immense and it is growing too quickly. You can have a general sense of many fields, a detailed knowledge of a few, and real expertise in only one or two. I speak here of people like myself, who are well-trained and have been in the business for a long time.

For a graduate student we must downsize the picture quite a bit. You are endeavoring to get through graduate school in a finite amount of time so that you can go off and live your life—get married, get a job, raise a family, and all of the rest. The only way to do this is to specialize drastically. While you are a graduate student, you are going to choose an area of study. Let us say that it is complex analysis. I assure you that you are not going to become expert in all of complex analysis while you are still a student. You are just going to learn a piece of it—extremal length, or the Bergman kernel, or value distribution theory, or conformal invariants, or approximation theory, or analytic capacity, or harmonic measure (and I mention these particularly because you, the reader, are likely not to have heard of any of them)—and burrow down deeply into that piece of the subject. You will learn about your thesis problem and about neighboring stuff that pertains directly to that thesis problem. You may need to learn a few more distant mathematical ideas—some notion from group theory, or differential equations, or harmonic analysis—and your thesis advisor will help you in those excursions. But, for the most part, you must specialize in order to get the job done.

One upshot of these considerations is that you will be frequently and painfully aware of all that you do not know. Not to worry. One of the main things that an Assistant Professor does in this discipline of ours is to tool up. That is the next main hurdle after completing the Ph.D. The new Assistant Professor will spend long hours reading, thinking, and discoursing about complex analysis, or whatever the chosen subject area is, just to become something of an expert. But you don't start out—fresh out of graduate school—as an expert. Nobody does. So adjust your expectations accordingly.

5.6. Why Does Everyone Else Appear to Be Succeeding?

Review Sections 5.2 and 5.5 and glance ahead to Section 8.7. Let me assure you that the query in the section title represents a syndrome that will affect you not only as a graduate student but also as you go on in the profession.

When you are a student, it will appear that

- Everyone else is breezing through their quals;
- Everyone else is getting a thesis advisor and a thesis problem with no effort and diving right into the work;
- Everyone else is making great headway on their thesis problem;
- Everyone else is solving their problem and writing a great thesis;
- Everyone else is getting lots of job offers and going on to great careers.

After you have graduated and you have been in the profession for a while, it will appear that

- Everyone else is getting invited to all of the big conferences;
- Everyone else has lots of research grants;
- Everyone else is winning Sloan Fellowships;
- Everyone else is getting invited to speak at the International Congress of Mathematicians;
- Everyone else is getting the plum jobs.

When you are a senior member of the profession, it will seem that

5.6. Why Does Everyone Else Succeed?

- Everyone else is getting elected to the National Academy of Sciences;
- Everyone else is getting a Chair Professorship at Harvard;
- Everyone else is winning the Wolf Prize;
- Everyone else is getting the Steele Prize.

And on it goes.

Ours is a competitive profession. We are all trained to be compulsive overachievers. We have spent our lives being praised for our little achievements and never finding any hurdle too great. We are accustomed to success and the accolades that go with it.

Once you get to the level of writing a thesis in mathematics, however, you are a member of a rather rarefied populace. *Everyone* that you know, *everyone* with whom you work, is very talented—probably about as talented as you. You are no longer being compared to the guys in your third-grade class who had IQ's comparable to their shoe size; you are being compared to the top 1% percent of the population.

There are certainly those among us—the future Fields Medalists and Bergman Prize holders and MacArthur Prize winners—who really do have an outsize share of prizes and encomia. You will see their names, photos, and achievements in the *Notices of the AMS* with tiresome regularity. They are our heroes and we can all strive to be like them. Most of us are not.

Most of us, if we are very lucky, win one or two awards in our lifetimes. These could be an MAA Award for Service to the Profession, or a Teaching Award, or it could be one of the big research prizes. The point I am trying to make is that most days, most of the time, you just go to work and do your stuff. You make incremental progress just like everyone else. You set reasonable goals for yourself and, eventually, you achieve them (in some form).

You may have been a child prodigy, but now you are just another mathematician. Your progress, achievements, and contributions are probably like everyone else's. On the other hand, mathematics is one of the finest and most erudite achievements of the human mind. The scholarly standard in the mathematics profession is one of the highest in the academic world. It is something to be proud of. And you are part of it.

Chapter 6

Moral Difficulties

6.1. Am I in Competition With My Fellow Graduate Students?

Well, you shouldn't be. But there are some issues of which you should be aware.

I have worked at certain universities where the beginning graduate students wouldn't talk to each other. They figured that the qualifying exams were graded on a curve and they didn't want to give each other any help. Such a situation is really a shame. For the most part, the quals are *not* graded on a curve. To pass a qual, a student must meet an objective standard. The faculty is neither frightened of flunking all of the students taking a given qual, nor of passing them all.

As indicated elsewhere in this book (Section 3.3), graduate students can give each other a great deal of help in preparing for the quals. They can study together, quiz each other, make each other sharp in internalizing the ideas and turning the theorems over in their minds. My view is that you should perceive the qualifying exam process as a group effort. Your class is working together to get past the first big hurdle of graduate school.[1]

What about the thesis? Surely there cannot be competition among graduate students at the thesis level. Can there? After all, everyone is

---

[1] When I was a student at Princeton, the custom was that the majority of graduate students would take the quals (which at Princeton were called the "Generals") at the end of the first year. Then we would celebrate by putting on an evening party for the entire department. Part of the fun was that we would write a satirical play ridiculing both the faculty and our fellow students, and we would perform it with great gusto at the party. Faculty and students alike came to the party with high expectations of a good time and some really bad jokes. Such was the camaraderie among the first-year graduate students in the early 1970's.

working for a different thesis advisor on a different problem. What is there to compete over? Unfortunately, this is not an accurate description of what can actually happen.

Suppose that one thesis advisor has several graduate students (I happen to know one particularly active advisor who has ten Ph.D. students right now!). Of course the grad students all compete for the advisor's time and attentions. It would only be human nature, and the natural course of things, for one or two students to emerge as the advisor's favorites. Perhaps s/he would give them more time or more fruitful ideas. Perhaps the advisor would take those special ones to a conference, but not the others. Perhaps s/he would do a better job of finding employment for the special ones.

In fact there is always a problem—built into the system—when a thesis advisor has several students getting their Ph.D.'s in the same year. The advisor will most likely be sending the same letter of recommendation about each student to every school to which the student applies. And it is almost unavoidable that one student among these will emerge as the strongest or most desirable—at every school! As a result—unfortunately—one student may get all of the offers and the other students may not get any.

This all sounds rather draconian, but if your thesis advisor is not a very clever letter-writer and politician, then this can really happen. It has happened to me with my own students. One year I had two really terrific students, but one had already done some significant publishing, had been invited to be a major speaker at some conferences, and was something of a celebrity. It was hard to hide the fact that he had several qualifications that the other student did not. As you can imagine, the one guy got all of the offers and the other got none.

There was a famous case at Harvard, about 25 years ago, of two students finishing with the same advisor. One of these, call him A, had worked very closely with the advisor. The other, call him B, had been quite independent. The advisor wrote a glowing letter about A, as he had a very good idea of what this student had done in his thesis, and thought very well of it. He wrote a rather vague letter about B since he only had a foggy notion of what that guy had done. Well, A got a good job and B really ended up at a terrible place, but it turned out that B was the real mathematician. He wrote 12 sensational papers in his first year out of school, set the world on fire, and ended up a tenured professor at one of the top math departments. Nobody is quite sure what ever happened to A.

In the year that I finished graduate school, there were 25 students getting their degrees in algebraic topology (and only one in analysis—namely me) at Princeton. There were only three algebraic topologists on the faculty, so they were the advisors to these 25 fresh Ph.D.'s. The custom at Princeton at that time was for the advisors to just shuffle us off to the plum jobs—the Moore Instructorships at MIT, the Dickson Instructorships at Chicago, the Peirce Instructorships at Harvard (see Sections 7.2 and 7.3). It was a halcyon time, but with 25 competing students there simply were not enough plum jobs to go around. So there was a very unpleasant meeting held for the 25 students and the three advisors in which people were simply *told* which jobs they could apply for and which schools they could consider. I know several people whose career paths were determined at that meeting, and not with a trajectory that they might have anticipated.

What I am telling you here is God's truth. It is a slice of life and much of it is completely out of your control. For the most part, you can avoid unpleasant and unproductive competition with your fellow students, and you should do so. It is always better to work together. If your thesis advisor has several students (yourself included) finishing in the same year, then sit down and talk to your advisor about what the parameters are and what your prospects are. You can have some influence over the situation and you should. After all, this is your life and your future that we are planning.

6.2. What Does it Mean to Be "Fired" by My Thesis Advisor?

Don't get chills. This doesn't happen very often and it probably won't happen to you, but I have actually fired two of my Ph.D. students. That is to say, I told them that I wouldn't work with them anymore, that they should either find another thesis advisor at Washington University (my institution) or else transfer to another program. [I was really thinking the latter, because if I wouldn't work with these guys, then nobody would.]

Both students, in fact, chose to transfer to another program, and I am happy to say that they are both doing well now, but they were not doing well at all at old Wash. U. The entire faculty had lost patience with both of them and I was the unfortunate "point man" who had to pull the trigger and get rid of them.

What was the problem? Had they kidnapped my dog, or failed to pay the right compliment on my new necktie, or slammed a door in my wife's face? No, nothing of the sort. There is really only one basic way that a graduate student can get fired, and that is to be unwilling to do the necessary work. In the case of one of these students, the issue was doing the work necessary to pass the Algebra Qual (he flunked it three times). In the other case, the issue was learning the subject area for the thesis.

In spite of the fact that a Ph.D. program is a long, tough haul, in spite of the fact that there can be many disappointments and missteps along the way, the people in your graduate program know how to get you through it.

And they will. But you have to be willing to do the work and to listen to the advice you are given. If you will not cooperate, then there can be trouble.

6.3. Academic Integrity

Most universities have a booklet or document about academic integrity. It is in your best interest to familiarize yourself with it. The document will define academic integrity, describe the common infractions (and the penalties for these), and discuss the judicial proceedings for handling miscreants. You will probably only rarely have to deal with academic integrity problems (plagiarism, cheating, etc.),[2] but it is good to know the parameters of your life.

When I signed on the dotted line to attend graduate school at Princeton, I had to make a separate attestation that I would adhere to Princeton's honor code. Their code is very strict and well-defined, and rather unusual. Students pledge when they are admitted that they will *turn in fellow students whom they observe cheating.*

I will now, just for the record, give an illustration of the Princeton honor code. For undergraduates at Princeton, the professor writes these words on the blackboard before each exam:

> I pledge my honor that I have neither given nor received information during this exam.

Each student must copy the words onto the exam paper and then sign the statement.[3] The instructor is *required* to leave the room during the exam. The instructor returns briefly, about 3/4 of the way through the test, to see whether there are any questions. The instructor leaves again and returns at the very end to collect the tests.

As an incoming graduate student at Princeton, I had to learn and agree to adhere to all parts of the honor code—even though it turned out that most of it could never apply to me. In the Princeton Math Department we had no grades and no exams; our only duties were to pass the quals and write a thesis, so most of this draconian folderol had no bearing on my life. But I was a sworn subscriber.

You couldn't cheat on the quals because the qual consisted of you locked in a room with three famous professors for three hours and they asked you

---

[2] If you encounter a cheating problem when you are a TA, then take it to the professor in charge of the course. Let him/her worry about it. That's why they pay professors the big bucks. When you are a faculty member, bring the matter to the Director of the Undergraduate Program or the Chairperson. There are legalities involved in cheating cases that are beyond most of us. You will need help.

[3] One wag has commented that the Princeton system forces a student who is planning to cheat also to lie.

6.3. Academic Integrity

questions orally. You answered them—in real time, on the fly—right there on the blackboard. How could you cheat on the quals?

Of course, many schools have written quals. And a student could, if determined to do so, cheat on a qual just as any student might cheat on a written exam. Be forewarned: The punishment for such misbehavior will be severe and forceful. The student may not be drummed out of the program, but will probably end up wishing s/he had been. Soon everyone in the department will know that X was the person who betrayed his/her colleagues. It will be the mistake of his/her (mathematical) life. A word to the wise should be sufficient. Your graduate math department ends up being like your extended family. Any infraction of this kind will be taken very personally and dealt with harshly. Don't even think about it.

Could a person instead cheat on the thesis? There are famous cases of students having their theses stolen—from their cars, their lockers, or even from their apartments. I had a friend in English at Princeton who kept the original draft of his thesis in a locked briefcase *with him at all times*. As he completed each chapter he put a copy in a safe deposit box and sent another copy to a (secret) remote address.

This type of neurotic behavior is not the norm in mathematics. Fortunately for us, mathematics is quite an open subject. We all talk to each other about what we are doing. We share ideas freely and expect to get useful counsel and advice in return. Your thesis advisor knows what you are up to and probably some of the other faculty do as well.[4] If, by some happenstance, somebody stole your thesis or tried to steal some of your ideas, everyone would know right away. The miscreant might be able to submit your ideas as a thesis at some remote, third-rate university. Oh well, there is hardly anything we can do about that. But, in the great marketplace of ideas, you can feel fairly confident that your ideas are safely yours *as long as you have shared them with responsible people who will stand up for you*.

Out in the real world, where people are faculty members at universities, there are various ways that they can cheat on their colleagues, their departments, and themselves. Sometimes a multilingual faculty member will translate a paper from some obscure language into English and then try to represent the work as original work—submitting it to a journal for publication as though s/he had created it and written it. Sometimes mathematicians will try to submit the same paper to more than one journal, hoping to pump up their publication lists with redundant works. Occasionally somebody will really try to abscond with another's ideas and represent

---

[4]This will be so either because you have been talking to other faculty, or else your thesis advisor has been talking to other faculty *about you and your work*. Mathematics really is a free and accessible subject.

them as his/her own, but for the reasons indicated in the last paragraph, this rarely works.

A good rule of thumb, when practicing mathematics, is to be generous. If someone gives you a good idea, or even the benefit of a stimulating conversation, then thank him/her. Offer the thanks in person, but also put a thanks in any paper that you may write. It costs you almost nothing to do so and your courtesy will be much appreciated. Everyone likes to know that his/her efforts have had a good effect and you will eliminate any possibility of resentment or ill will by simply being forthright and generous to begin with. And you can then expect other mathematicians to behave in the same way toward you. Mathematics is a process, and we are all in it together.

Fortunately for us, most mathematicians are quite honest most of the time. You have little to worry about (*vis à vis* academic integrity) while you are a graduate student and you have many other things on which to concentrate your efforts.[5]

6.4. Intimacy with Members of the Mathematics Department

Of course you should socialize with your fellow graduate students. Date your fellow graduate students. Go further than that if everyone is consensual. Heck, marry them if you so desire. But if you want to get intimate with the faculty, the staff, or the undergraduates, then you are on dangerous ground.

One of the few ways that a faculty member can lose tenure is through moral turpitude. In particular, a faculty member having sex with a graduate student qualifies—just because it can either be caused by or be the cause of an abuse of power (i.e., it could entail sexual harassment). The trouble is that working closely with a thesis advisor often leads to a tight personal relationship. When you are both on the brink of a big discovery, emotions can run high. Anything can happen. Sometimes graduate students and their thesis advisors do get involved, but it is treacherous territory. The emotional relationship can impinge on the professional relationship in a number of unpleasant ways. I only offer this information as a caution.

Likewise, it can be very tricky for a graduate student to get intimately involved with an undergraduate or with a member of the staff. There are usually no explicit rules against it (as long as nobody is underage), but the university must be concerned about sexual harassment and about abuse of

---

[5]I was once chatting with one of my undergraduates at UCLA about a problem he was working on at some company at which he was employed. At one point he offered to share his calculations with me. When I assented, he handed me a sheaf of mathematics written on special paper with an inscription across the top that said that the paper would self-destruct if I attempted to photocopy it. I have been forever grateful that academic mathematics does not work that way.

6.4. Intimacy with Colleagues

power. If a graduate student at the university breaches the law, then the institution is at risk as well.

Several years ago the University of Minnesota instituted a policy prohibiting a full Professor from having coffee with an Associate Professor, an Associate Professor from having ice cream with an Assistant Professor, and so forth and so on. This came about because there had been such difficulty with sexual harassment issues. These days, people have been hypersensitized to the problem and it is all too easy to file charges. It is also miserably difficult to defend yourself against such charges. If you get in trouble for sexual harrassment at one university, then the onus will likely follow you around all of your life. You just don't want to get tangled up in this nexus.

As Woody Allen said when he got involved with an underaged woman, "The heart wants what it wants." True enough. As adults we have freedom of choice in all that we do, but sometimes there is a price to pay and you should be aware of the risks.

Part 4

Post-Graduate-School Existence

Chapter 7

Life After the Thesis

7.1. Should I Publish My Thesis?

Of course you should. It contains your original ideas and the fruits of your protracted labor, it represents a monumental amount of work, and you should publish it. I know of situations in which a Ph.D. candidate published as many as three solid papers from his Ph.D. thesis. Thus this person's reputation was established instantly.[1]

This is another good time to get advice from your thesis advisor. Should you publish your results in one paper or several? How do you reduce the rather florid and discursive verbiage of a thesis to the more telegraphic style that is suitable for a published paper? How long should it be? How should the references be formatted? Whom should you thank? To what journal should you submit? The book [KRA2] contains extensive detail on how to write up a paper and on the process of submitting the paper to a journal. You can submit your paper to only one journal at a time, so exercise some care in choosing the correct forum for your work. Also note that each journal has specific style criteria that you should follow. There will be instructions as to how many copies to submit, whether you should include a disc, and additional particulars. Read the "Instructions to Authors" page that appears in most issues of your target journal.

---

[1] When my colleague Mitch Taibleson earned his Ph.D. over thirty-five years ago, his thesis advisor (also my thesis advisor) felt that Mitch's thesis was very important. He advised, and Mitch tried, to get the very long thesis published in one piece in one of the preeminent analysis journals. For a variety of political reasons, this didn't work; the paper was rejected. Zygmund then advised Taibleson as follows: "Break your thesis up into ten papers and send them to ten different journals. You will thereby establish your reputation immediately. You can worry about being ethical later."

It sometimes happens—not very often—that the thesis advisor will feel that your thesis contains significant ideas that are mostly due to him/her. So, if there is to be a published paper emanating from the work, then the advisor's name should be on it. Some thesis advisors (especially those from Eastern European countries) will even go so far as to insist that their name go first—even if that violates the usual custom of alphabetical order for authors' names.

This is a tricky situation, and I can only offer this advice: Everyone will know that this is your thesis and that your advisor played a decisive role in making it happen. There is no way to hide that fact. Your thesis advisor's name on the paper doesn't really change anything; it simply validates an established fact. My advice, then, is (if the question of your thesis advisor's name being added to your published work comes up) just to swallow your pride and go along with having your advisor as a coauthor. The alternative could be quite ugly and result in no published paper at all, which would really be too bad.

Whether you decide to pursue a career in academics, business, industry, or the government, it is always a good thing to include some publications on your *vita* or résumé. If you don't publish your thesis, then the results could become superseded or subsumed by work of others. You may lose your claim to the results. So I encourage you to plan to publish your thesis. If you go on in academic life, you will want this item on your publication list (when it comes time for the tenure decision—see Sections 7.2, 7.6, 8.4, and 8.6); people will raise their eyebrows if it is not there.

If you want to establish yourself in the academic world and be a good and strong candidate for tenure, then it is generally a good idea to move away from your thesis and to develop in new directions as soon as you can. Then you really *will* be on your own and your papers will be entirely yours. I know mathematicians—some of them excellent—who published extensively with their thesis advisors in the early part of their careers. Let me stress that this was not done under duress: the young mathematician and the thesis advisor were great friends and had many common interests; they really *wanted* to follow this research path together. Unfortunately, this worked against the younger coauthor when tenure time came. As you can imagine, the tenure committee and the Dean had trouble sorting out how much of the joint work was the contribution of the (young) tenure candidate and how much was the brainchild of the more experienced (older) thesis advisor. I only mention these exigencies so that you can be aware of complications that may arise.

7.2. What is a Tenure-Track Job?

There are academic jobs and nonacademic jobs. Academic jobs come in two flavors: tenure-track and non-tenure-track. Let us begin by discussing that distinction.

Tenure is a grand intellectual tradition by means of which security is provided to academics. The theory is that they need this security so that they can explore daring new ideas without having to worry about societal pressures and bugaboos or being accountable for their time. Tenure protects a mathematician, for instance, so that s/he can invest time in changing fields or exploring a potentially unproductive byway. Qualifying for tenure is an arduous process. The candidate has a six-year probationary period[2] to establish a solid profile as a teacher, a department member, and a scholar. When the tenure decision is made, letters are solicited from experts all over the world.[3] A teaching dossier is created and the candidate's contributions to the departmental weal are analyzed (see this section and Section 7.6 for the chapter and verse of the tenure process).

A departmental meeting is held at which the candidate's qualifications are scrutinized and discussed in detail. A vote is taken. If the mandate from the department is a strong one, then the case is passed on to the Dean. The Dean and the Promotion/Tenure Committee analyze the case, paying attention to the three criteria (research, teaching, service) already indicated. They conduct a critical discussion and hold a vote. If their vote is strong, then they pass the case up to the Provost. After an analysis at the Provost's level,[4] the case finally goes to the Chancellor and the Board of Trustees.

If at any stage in this process the vote taken is not a strong mandate, then the case will probably die. Of course there are various appeals processes of which the candidate may avail himself/herself. Sometimes negative cases are reversed. But this is the gist of the process.

The reason that such care is taken is that tenure is for life. It is difficult, complicated, and agonizing for a university to divest itself of a tenured

---

[2]This, and other aspects of the tenure process, is mandated by the American Association of University Professors (AAUP).

[3]This situation must be tempered according to circumstances. A top research university will certainly seek outside letters in support of a tenure or promotion case. A college whose primary mission is teaching, and at which research is not a large component of the faculty evaluation process, will probably not seek outside letters (at least not very many). After all, it really does not make any sense to ask a big expert at Harvard about the scholarly reputation of someone whose primary duty is to teach undergraduates and who has published very little.

[4]It is worth mentioning that the different levels of the review process will place different emphases on the different components of the case. For example, often the Provost will reason that the scientific merits of the case were treated thoroughly at the Departmental and Dean's levels. So the Provost will concentrate on teaching, service, and other matters.

professor, so the institution is particularly careful at the outset in deciding who will receive tenure.

The *tenure track* is the seven-year period in which an Assistant Professor is hired, assessed, and then, finally, considered for tenure. Obviously all of us, eventually, want to be hired into the tenure track at a good school. As we will see in the discussion below, if is often advantageous to begin one's career with a *non*-tenure-track Instructorship or Postdoc.

There is room at a research university for non-tenure-track positions. Fresh Ph.D.'s relish the opportunity to go to a recognized research center for a few years to work with established experts in their field. Instructorships (sometimes called "Postdocs") have been created for that purpose. Such a job lasts two or three years and it is terminal. There is no possibility of tenure in such a job. At the end of the Instructorship the candidate will most likely seek a tenure-track job at another institution.

Many schools, especially the elite private ones, have endowed research Instructorships. It is a real honor to land one of those. MIT has Moore Instructorships, Harvard has Peirce Instructorships, Chicago has Dickson Instructorships, and UCLA has Hedrick Instructorships; many others exist. Often these Instructorships have a very nice salary, a reduced teaching load, a travel allowance, and other perks. Of course the most important benefit is that you get to work with the best and most active faculty for a few years and to really jump-start your research career. Again, such Instructorships last two to three years. They are terminal and *not* part of the tenure track.

Some new Ph.D.'s, especially those who already have a family, will seek a tenure-track position as their first job—straight out of school. They do this because they are seeking the security of tenure and do not feel that they have the discretion to spend two or three years in a Postdoctoral position.

These days, the research universities generally will not hire fresh Ph.D.'s directly into tenure-track positions; they prefer only to consider experienced people for the tenure track. From their point of view, a fresh Ph.D. is a candidate only for an Instructorship.[5] The top Liberal Arts Colleges and many of the comprehensive universities also prefer to hire tenure-track faculty who have had a couple of years experience and have established a solid track record elsewhere.

---

[5]Such a person can, of course, proceed into an Assistant Professorship *after* the Instructorship. Then the tenure-track process proceeds as usual.

7.2. What is a Tenure-Track Job?

Four-year colleges, where the primary faculty duty is to teach (and where research plays a secondary role), are a different type of situation.[6] They generally do not have research Instructorships. Most of their full-time jobs will be tenure-track. So if you seek a tenure-track job straight out of school, then you should pay particular attention to schools of this kind. *Comprehensive universities*, which are not elite research institutions but are where most of the college teaching in this country takes place, are also places where you can get a tenure-track job straight out of graduate school (see below). The publication *Employment Information in the Mathematical Sciences* (*EIMS*), discussed in what follows, is the primary resource for academic job information in mathematics. The *Chronicle of Higher Education* is one of the main venues in which teaching institutions advertise their openings. Advertisements also appear in the *Notices of the AMS*. The AMS web site is also a good source of job information.

The tenure process at a four-year college is similar to what was described above, but a greater emphasis in the tenure decision is placed on teaching (and a correspondingly lesser emphasis is placed on research). Liberal Arts and other schools that stress teaching and student interaction have special and detailed methods for evaluating teaching. They want to be *sure* that they are tenuring people who know how to teach and who are committed to the teaching process.

Comprehensive universities (in many cases these are the former normal schools or teacher-training colleges) do not have research programs or research-oriented faculty. They are often divided up into colleges—a business school, an engineering school, and so forth. A comprehensive university will grant Master's degrees, but usually not Ph.D.'s. These institutions have plenty of students and plenty of jobs and the young Ph.D. seeking a position had better be aware of them. They do not have Instructorships or Postdocs; they hire many mathematicians straight out of graduate school into tenure-track positions. [Comprehensive universities do have nontenured people whom they hire on one-year contracts. It is possible to get some teaching experience in such a position.] You will certainly see their advertisements, in great number, in the *Chronicle of Higher Education*.

Of course there are also nonacademic jobs, and plenty of them (see Section 7.4). Ph.D.-level mathematicians are in great demand these days.

---

[6] Many of these four-year colleges are what are commonly known as "Liberal Arts Colleges". These are institutions devoted to undergraduate education at the highest level. Harvard College (part of Harvard University), Swarthmore, Haverford, Macalester, and Williams are among the best of these. There are many hundreds throughout the country. Some four-year colleges specialize. Harvey Mudd and Rose-Hulman are very fine engineering schools. They could not properly be called "Liberal Arts Colleges". Some others specialize in the performing arts, or fine arts, or other particular areas of the academic enterprise. They are certainly a large part of what makes the American educational system rich and diverse.

Government agencies, private high-tech firms, scientific labs, financial firms, businesses, and manufacturers of all sorts need mathematical expertise. Mathematicians have developed a strong reputation as people who can think critically, solve problems, and analyze difficult situations. The opportunities are many. More will be said below about this matter.

7.3. Looking for a Job

Your thesis advisor can, and should, play a central role in getting you a job.[7] I earned my Ph.D. in ancient times. My advisor just picked up the phone and landed me a tenure-track job at a good school. There were no applications, no letters, no nothing. It was all just arranged with a single phone call.

This type of politicking is virtually unheard of today. First of all, there are government rules such as the Affirmative Action Program that are designed to discourage such behavior (to wit, it has been felt that the "good old boy" network was intrinsically unfair, and tended to exclude women and other underrepresented groups). Secondly, jobs are not quite as plentiful these days (although there are lots of jobs, and many retirements coming up in the next ten years, so don't worry).

For the most part, you must go through the formality of submitting job applications. This includes completing an AMS Cover Sheet[8] (available from the American Mathematical Society web site http://www.ams.org or in issues of the *Notices of the AMS*), putting together a *Curriculum Vitae* (see [KRA2] for some advice on how to write your *vita*), getting letters of recommendation (usually three), and writing a Teaching Statement[9] and a Research Statement.[10] Your job application should contain a complete

---

[7]This statement may seem obvious, but it is not. For example, in the rather distinguished School of Architecture at my own university, the faculty provide the students no help in finding jobs.

[8]This is an innovation introduced several years ago by the American Mathematical Society. From the point of view of the potential employers (i.e., the math departments), it is a good idea, because it guarantees that every job-applicant's dossier will contain basic information: school of Ph.D., year of Ph.D., subject area, name of thesis advisor, etc. I encourage you to include this standardized Cover Sheet in your applications to colleges and universities.

[9]It has come about in the past five or ten years that a young job applicant in mathematics is expected to provide a Teaching Statement. This is an essay of about two or three pages describing your philosophy and experience of teaching. To be honest with you, it is rather difficult to say anything interesting and new about teaching—given the level and depth of experience you have had and given the amount of ink that has already been spilled on the topic, but you *must* do this, and you must do a very good job of it. Everyone who reads your dossier will read your Teaching Statement and will derive some sense of who you are from what you write, so take this matter very seriously.

[10]A young mathematician applying for an academic job is also often expected to supply a Research Statement. This brief (two- to three-page) document will describe the candidate's research achievements and future plans for additional work. It should be written so that anyone

7.3. Looking for a Job

teaching dossier.[11] Let me assure you that the reputation, the enthusiasm, and the energy of your thesis advisor can make a huge difference in what kind of job you will get. You should think about this matter a little when you select a thesis advisor.

Let me stress, once again, that Liberal Arts schools, comprehensive universities, and four-year colleges place a greater stress on teaching than on research; some of these institutions place a *much* greater stress. So if your job application has a ten-page Research Statement and a brief and lackluster Teaching Statement, then it will *not* be given serious consideration at such a school. You should consider *tailoring* your application for the type of school to which you are applying. People on the hiring committees at teaching schools tell me that the only thing they really look at is the Teaching Statement. And—since it is so difficult for you, as a neophyte in the business, to say anything really new about teaching—what they are looking for is a sense that you just love to teach, that you are going to throw yourself into the work with passion and commitment. That's what you want your Teaching Statement to demonstrate.

As noted, you will need at least three letters of recommendation in your job-application dossier. One of these should be from your thesis advisor. This will be a long and detailed letter about you and your work. Although your advisor will no doubt say some things about your teaching abilities, someone else usually writes a separate letter about your teaching; this letter will most likely be from the faculty person who supervises the TAs.[12] The third letter is usually from some other faculty member who is familiar with your thesis work. This is yet another reason why you should always be talking to people about your stuff. It is quite common for a finishing graduate student who is looking for a job to just knock on somebody's door (cold) and say, "I need a job letter. Let me tell you a little bit about my thesis and then you can write one for me." Quite common, but it doesn't work

---

assessing the dossier can get a sense of what this candidate is about—mathematically speaking. If you have a research grant—either from the NSF or from your university—certainly include that information in your Research Statement.

[11]I have been in the profession for thirty years. I authored the book *How to Teach Mathematics* and have won a major teaching award. My reputation as a teacher is well known; I probably do not need a teaching dossier. If you are just beginning as a mathematician, then your reputation has yet to be established. You *must* have a teaching dossier. It will lay out explicitly what courses or recitation sections you have taught, what your duties were, and what other teaching activities you have engaged in. Have you worked in an instructional lab? Do you have experience teaching with computers? Are you acquainted with "reform" teaching methods (see [KRA1])? Have you taught any statistics? Have you tutored? Don't be modest here. Tell the person evaluating your dossier everything there is to know about your qualifications as a teacher. Include numerical data from teaching evaluations if it is available. If you have won any teaching awards, definitely say so.

[12]It is best if you can get this person—or *someone*—to actually watch you teach and then report in his/her letter on what they saw. This makes for an especially credible and compelling teaching recommendation.

very well. The resulting letter is usually terse and with little detail. Much better is to have developed a relationship with some faculty member other than your advisor. Then the person can really comment in detail about you and your work.

I must say once again that a school whose primary mission is teaching is probably not interested in long disquisitions about your theorem-proving abilities. Of course such a school wants to know that you are a solid, well-trained scholar who has written a fine thesis, but your job application letters of recommendation to such a school should make quite a production out of your presence as a teacher and departmental member. They should of course say some nice things about your research, but that should not be the emphasis. To repeat, it is good planning to tailor your application to the type of school to which you are applying.

What really makes a terrific impression in your job application dossier—at least to a research institution—is getting some letters from faculty *outside* of your university. I often invite colloquium speakers who would be of interest to my graduate students and arrange for the students to spend some time with the guests. This frequently results in the outside faculty member becoming familiar with my student's work, perhaps even reading the thesis. Then this person can write a really nice letter on behalf of my student. More than one of my Ph.D. students has written a paper with a mathematician at another university—even before graduation.

An important part of your job application is your cover letter. It gives the person reviewing your dossier a quick idea of who you are and whether the reviewer wants to read on. At the very least, you should give your name, address, *e*-mail address, areas of interest, and a couple of words about your teaching experience. For a job at a research university, your cover letter can be quite brief (about one third of a page). For a job at a comprehensive university or Liberal Arts College, your cover letter should expand a bit on your career goals, your teaching philosophy, and your views of departmental service. Take no more than a page to complete this précis.

In spite of what I have been saying about the crucial role of the thesis advisor in getting you a job, you can do many things to help yourself. For one thing, sign up for the job interview activities that the AMS (American Mathematical Society) conducts at the annual joint AMS/MAA meeting in January.[13] Many schools like to meet a broad cross-section of candidates at the Employment Center and then winnow the list down to those few whom they will fly to campus for a more extensive meeting. You sign up in advance (usually in October, so be forewarned!) for this activity and

---

[13]This used to be known as the Employment Register. It is now called the Employment Center. It is sponsored by the AMS, the MAA, and SIAM.

7.3. Looking for a Job

pay a modest fee. Part of your application is to provide a résumé and to list several schools by which you would particularly like to be interviewed. Conversely, the schools list candidates to whom they would particularly like to speak (after having read the résumés) and then the AMS does a computer match. Depending on an applicant's qualifications, he or she may get a few or several interviews. Five is about average, but some candidates have had as many as twenty-three. If you are particularly attractive, schools will contact you independently of the computer-matching process. One of my recent students ended up with about thirteen interviews at the January AMS/MAA meeting, only five of them arranged through the offices of the AMS.

There is also a free-form component to the Employment Center activities. Many institutions (especially research universities) and many private-sector employers and government agencies choose not to participate in the computer-matching procedures described in the preceding paragraph. Instead, the interviewers set up a table and meet with job candidates who come by *at the meeting*—without any appointment. Candidates can also arrange interviews in advance with the free-form participants (*not* using the computer-matching service). In recent years, the number of these "free-form" interviews has exceeded the number of computer-matched interviews.

Most of the interviewers at the AMS Employment Center are academic, but a few are from Aerospace Corporation or some of the other high-tech employers. The National Security Agency often interviews at the Employment Center. Almost all of the jobs being offered are for holders of the Ph.D. In January of 2002, there were 63 employers conducting interviews and 246 job-seekers. A total of 1797 interviews were conducted.

Here are some special points to note about schools that interview and schools that do not:

- As previously noted, it has been the case for many years that the top fifty math departments prefer to hire new Ph.D.'s as (temporary) two-year or three-year Instructors. An interview is generally not required for such a position.

- For tenure-track jobs, the top fifty departments prefer to hire someone who has been out for a while and had a chance to establish an independent reputation; an interview will definitely be required for such a tenure-track job.[14] The candidate is somebody who will be considered for tenure in a few or several years and may become

---

[14]Sometimes, especially with departments that do not have a lot of discretionary money to fly in candidates, the interview will be conducted by telephone. There may be a conference call that includes the Chairperson, members of the hiring committee, and the candidate.

a permanent member of the faculty. Of course the hiring department (and the Dean as well) will want to meet the candidate and determine his/her suitability.

- Colleges that are not research-oriented generally do not have Instructorships. Most of their full-time jobs are tenure-track, although such colleges do generally have some "visiting positions" or other types of temporary positions. Four-year colleges and Liberal Arts colleges and comprehensive universities will often want to interview all of the job candidates on their "short list".

- Many, if not most, of your interviews at the January AMS/MAA meeting will be for tenure-track jobs. The computer-matched interviews will be primarily by four-year colleges, Liberal Arts colleges, and comprehensive universities. The free-form interviews will tend to be with research universities, and some will be with industrial and government employers.

The track record of offers actually tendered as a result of interviews in the Employment Center at the AMS meeting is quite strong. This is a process that really works and I encourage you to participate.

Overall, how many job applications should you submit (by mail)? When I was a graduate student—nearly thirty years ago—the students in the very top, elite programs often obtained jobs through their advisors' connections; there was not much applying. Those who did send in applications submitted ten or twenty. That was sufficient.

Now we live in a different age. The mathematics world has grown and diversified. Because of complicated global politics, a number of foreign mathematicians—many of them well-established—seek jobs in this country. The process is more competitive and more people are on the market.[15] In the search for academic jobs, it is quite common to submit 100 applications. Because of computers and word processors, and because at many schools you can apply on the internet, it is really not so difficult to apply to a great many schools. Most of your letter-writers will only write one letter about you, so it is no extra trouble for them.

Having said all of this, I must note that the math departments offering positions are feeling quite overwhelmed these days. When there are 500 or more applicants for two positions, it is difficult to conduct interviews and to make rational decisions. Many departments now ask an applicant whether the candidate knows anyone in the department, or who his/her contacts might be. If the answer is the empty set, then the application gets tossed

---

[15]Of course it should be noted that most of the foreign mathematicians will be established members of the field and will not be seeking the same jobs that a fresh Ph.D. will be seeking.

7.3. Looking for a Job

aside. Your time might be better spent making contacts and identifying job opportunities than engaging in the shotgun approach to the job-seeking process. The Employment Center at the January AMS/MAA meeting is certainly a good vehicle for this process.

If your application is successful, then you will be hearing from some schools. For an Instructorship or Postdoc, you will get an *e*-mail, a phone call, or a letter (most likely you will get all three) from the Chairperson of the Department to which you applied.[16] The Chairperson will offer you the job, describe the duties, salary, and benefits, and express the fervent wish that you will accept. You will be given a period of time—ranging from a day to a week—to consider the offer. Be forthright in your dealings with the Chairperson. If s/he wants an answer within 24 hours, and if you need more time (and can give a good reason), then say so. You will probably get a reasonable response.

You may be so excited to have a real job that you are tempted to accept your offer right over the phone. There is no reason to do so and plenty of reason not to. First of all, if one school is offering you a job, then probably others will also. You want to have a choice. Secondly, you will want to consult your thesis advisor, your spouse, your significant other, or perhaps your parents. Finally, you should just take some time to mull things over. You may want to contact some other schools, point out that you have an offer pending, and ask whether they have anything cooking for you. There is no reason, nor any need, to appear excessively anxious. They will give you time to think about the matter and you should take that time.[17] Think about whether the salary is appropriate, whether the duties seem reasonable, whether the institution is located in a place where you actually want to live and/or settle down. Do you know any people at this new place (or does your advisor)? Do you like them? There is plenty to consider and you should do some homework before accepting the exciting new offer that you now have on your plate.

When it is time to accept the job, you can use phone or *e*-mail, but you *must* follow up with a formal, written letter of acceptance. There will be certain formalities about health insurance and other benefits. The departmental Office Manager will be in touch about these details. Sometimes the

---

[16]It is also common to get a preliminary phone call in which you are told you are on the short list and you are asked whether you are still interested and still available. Unless you have already accepted a job elsewhere, a good rule of thumb is to answer "yes".

[17]When Salomon Bochner was Chairperson at Rice University, he was famous for phoning up young mathematicians with an offer to which he demanded a response right on the spot. If the candidate was not willing to play ball, then Bochner would just hang up the phone. Bang! AAUP rules, Affirmative Action, and just general good manners prevent people from behaving in this manner today.

hiring department will help with your relocation expenses and you can ask about that.

If you are instead applying for tenure-track jobs, then the process is just a tad more complicated. The schools doing the hiring will certainly want to interview you. There may be a preliminary interview at the AMS/MAA/SIAM Employment Center at the January national meeting. If that is successful, then they will fly you to their campus so that you can meet the whole faculty. You may be asked to give one or more sample lectures, showcasing both your mathematical credentials and your teaching credentials. Whereas hiring decisions for Instructors are often decided just by a committee within the math department, the hiring decision for a tenure-track job is usually made by the entire department. The Dean may also be involved. So it is liable to take a bit longer.

Eventually you will hear from the Chairperson, who will offer you the job (usually by phone) or not offer the job (in which case you will probably be notified by letter). You will be given about a week to consider the offer. *Definitely take the allotted time and discuss the matter with your mentors, your family, and/or significant others.* Then you can respond by phone or e-mail, but back it up with a formal, written letter of acceptance. You will hear from the departmental Office Manager about health insurance and other benefits. Be sure to ask whether the department will help with your relocation expenses!

Most people who earn their Ph.D.'s at a top research university expect (at least subconsciously) to end up at a similar place for their careers. Depending on economics and other societal factors at the time you get your Ph.D., this may not be an immediately realistic expectation. Typical figures (obtained from AMS data files) are these:

- In recent years, about 20% of new Ph.D.'s obtain first positions at Group I or Group II schools. Most of these positions are Postdocs.

- Fewer than 30% of new Ph.D.'s obtain their first position at a doctoral-granting institution.

Of course the second of these bulleted items means that at least 70% of fresh Ph.D.'s are going to comprehensive universities, Liberal Arts colleges, or teaching schools. This figure has been decreasing in recent years because of NSF programs to increase the number of Postdocs nationwide.

Be aware that the teaching colleges (e.g., the Liberal Arts colleges) and comprehensive universities are much less susceptible than are research institutions to the vagaries of the market. Teaching schools always have jobs— lots of them. The Liberal Arts colleges are mostly private, do not depend on the ups and downs of public funding, and do not go through wild swings of growth and decline. The comprehensive universities are the guts of most state higher education programs. They will always be with us and always have jobs. Look in the *Chronicle of Higher Education* classified ads to see the vast array of schools (about 2800 in this country) and the vast panorama of jobs that are available. Be aware of the ever-changing job listings on the AMS web site (see `http://www.ams.org/eims/eims-search.html`) and read *EIMS* (*Employment Information in the Mathematical Sciences*).[18] *But be sure to talk to people.* Talk to the junior faculty in your department. They have been on the job market recently; they have a good sense of how things are and what the job market is like. Talk to several different faculty at all levels; they will all have a different take on the job situation.

Many, many jobs are advertised over the internet or by *e*-mail. I get *e*-mail messages every day announcing new openings and positions—and these are not jobs for me, but instead are jobs for those within a few years of the Ph.D. Despite all of the new rules and regulations, many jobs are obtained through personal contacts, so develop your own personal network and use it.

The vast majority of advertised academic positions in mathematics each year are for young people—mathematicians who are five years or less from the Ph.D. To be sure, there are always openings for Deans or Chairpersons or Chair Professors; occasionally, ads for a senior mathematician to head up a program appear, but Deans prefer to hire at the lower rungs—this means fewer entitlements and at a considerably reduced salary level. An interesting feature of mathematical life these days is that a great many (about thirty) institutions have federal grants to enhance their graduate programs. These provide, among other things, funds for Instructorships. As a result, there are over two hundred *more* Instructorships (Postdocs) in mathematics than there were five years ago. Thus these federal programs provide many more beginning mathematicians with a head start in the profession.

7.4. Nonacademic Jobs

Even when I was a child, it was common wisdom that if your education was in mathematics, then the only job you could get was as a teacher. Today nothing could be further from the truth. So many parts of life—and not just the technical aspects—have become analytical that there is a huge demand for mathematicians. [Just as an instance, the L'Oreal cosmetic firm has a huge research operation that employs mathematicians who are creating a mathematical model of the face.] And while it is true that there was a time

---

[18]This information is available both in hard-copy form and on the AMS web site. The latter version is updated each week. If you register online, then EIMS will send you *e*-mail notifying you when new positions are available.

when employers were chary of hiring Ph.D.'s (because a Ph.D. is an egghead who won't do what s/he is told), nowadays math Ph.D.'s have a high profile and many positions for Ph.D. mathematicians remain unfilled for lack of candidates. Here are just a few of the opportunities in today's society for a mathematician with a Ph.D.:

- **The National Security Agency.** This is the largest employer of mathematicians in the world. It deals with a number of aspects of government work, much of it classified. One of the biggest areas of activity at NSA is cryptoanalysis. The NSA also does a fair amount of statistics as part of the "capturing" of signals.

- **The Institute for Defense Analyses.** Located in Alexandria, Princeton, and San Diego, this government agency treats various national security issues.

- **The RAND Corporation.** RAND is a think tank in Santa Monica that does a lot of consulting for the government. John Forbes Nash used to work at RAND and John von Neumann consulted there.

- **The Microsoft Mathematics Unit.** Microsoft has its own mathematical think tank and it also employs mathematicians in research and development.

- **Daniel Wagner Associates.** Another large employer of Ph.D. mathematicians, founded and run by mathematicians. The company does a lot of consulting for the government and for industry.

- **Hughes Aircraft and The Aerospace Corporation.** These are but two of the many high-tech companies that have long employed Ph.D.'s from UCLA and other universities. Aerospace has interviewed at the AMS Employment Center.

- **Mitre Corporation.** Founded by John Kemeny and others, this company concerns itself with a variety of mathematical questions arising from tech sector activities.

- **Wall Street.** Ever since the advent of the Black/Scholes option pricing scheme, and even before, stochastic integrals have played a prominent role in investment strategies. Mathematicians are in great demand here.

- **Companies ranging from General Electric to Texas Instruments to Sylvania to Hewlett-Packard.** The high-tech industry has use for mathematicians in software verification, in microchip design, in control theory and systems development, in operating

7.4. Nonacademic Jobs

system design and implementation, in general research and development, and in many other aspects of technical work. Microsoft employs mathematicians on its staff. So does SONY.

- **The Genome Project.** The mapping of the human gene, and the more recent area of proteomics, makes great demands both in mathematics and statistics.
- **The Actuarial Industry.** The analysis of insurance data and the design of annuities and amortization plans is a big industry. Actuaries are in great demand and mathematical training is the optimal preparation for this profession.
- **Medical Research.** From plastic surgery to psychology to medical imaging, there is great demand for mathematicians and statisticians. The math department at Washington University (my institution) has sent many of its Ph.D.'s to work in research groups at the Wash. U. Medical School. These days, medical schools have a great need for people with analytical skills and they are willing to retrain people to fit their programs.
- **Los Alamos, Oak Ridge, and Other National Labs.** These government research sites engage in many interesting long-range projects. The Ginsparg project for the electronic archiving and publication of scientific papers was initiated at Los Alamos and has received considerable attention. In fact, Ginsparg recently won a MacArthur Prize.
- **The National Census Bureau.** The collection and analysis of data for the census is highly mathematical. There are a number of cutting-edge ideas being developed to analyze these masses of data.
- **Systems Analysis.** Many Ph.D. mathematicians work as systems managers for large LANs and other industrial and academic computer systems.
- **Pharmacokinetics.** The question of validation of the effectiveness of generic drugs is one of many pharmacokinetic questions that has demanded great sophistication in statistical analysis.
- **The Navy.** The Head of the Office of Naval Research has attributed the American supremacy in navigation and radar detection systems to American mathematicians.
- **Data Mining.** This is an entirely new research area for developing techniques to analyze the billion-piece data sets that arise from genetic and other areas of medical research. Applications also exist in meteorology, cosmology, and other parts of analytical science. The ground has barely been broken in this vast new area, yet it

has touched on almost every part of technical work, ranging from government offices to industrial and academic research labs.

- **The Automotive Industry.** The low-emissions carburation system in the Volvo was designed by mathematicians. The computer-driven body-design software (called *SLIP*) that is used by many automobile manufacturers was developed by a mathematician at Washington University.

- **The Social Security Administration.** One of my classmates from graduate school is a high-ranking administrator at the Social Security Administration.

- **The National Aeronautics and Space Administration (NASA).** Not surprisingly, the space program uses a considerable amount of mathematics—in navigational problems, in vehicle design, in software verification, in the design and construction of space stations, and in the resolution of graphic images.

This list could go on at length. You should understand that the mathematical life can take many forms and can be stimulating and exciting at many levels. There are lots of things you can do besides teach calculus. Keep this in mind when it comes time to look for a job.

We conclude with a caveat. Most thesis advisors are quite knowledgeable about the academic life. They can advise and help you in detail about getting an academic job—especially at the university level. Unfortunately, most are much less well-informed about nonacademic jobs. You may be able to glean useful information about industrial jobs from faculty in engineering, computer science, and applied mathematics. Don't be afraid to ask.

Quite frankly, the value system and *modus operandi* in the nonacademic world is very different from that in the academic world. As Bernard Beauzamy, CEO of Société de Calcul Mathématique, SA (SCM for short), puts it in [BEA]:

> Most current mathematical research, since the 60's, is devoted to fancy situations: it brings solutions which nobody understands to questions nobody asked. Nevertheless, those who bring these solutions are called "distinguished" by the academic community. This word by itself gives a measure of the social distance: real life mathematics do not require distinguished mathematicians. On the contrary, it requires barbarians: people willing to fight, to conquer, to build, to understand, with no predetermined idea about which tool should be used.
>
> . . .

> Many young people write to us: they apply for a job at SCM. They write the following way: "I am qualified for a mathematician's job, since I have studied optimization, numerical techniques, this and that software, and so on." That's nice, of course, but this does not meet my concern. My concern is, primarily, to find people who are able and willing to discuss with our clients, trying to understand what they mean and what they want. This requires diplomacy, persistence, sense of contact, and many other human qualities. This is a matter of personality, which is not taught, never at school nor at the university, and which is much harder to acquire than any technical knowledge about, say, the simplex algorithm.

If you choose to seek nonacademic employment, then you will have to provide more of the initiative. Your advisor will still write a good letter[19] for you and provide a lot of encouragement, but s/he will likely have far fewer contacts and will be less familiar with the process. It is also the case that many faculty at research universities are relatively unfamiliar with the job situation (as well as the requirements and expectations) at a Liberal Arts college or comprehensive university. Your Ph.D. institution probably has many visitors each year (people on sabbatical from a variety of institutions, some of them Liberal Arts and comprehensive). Talking to them is one useful and convenient means for acquainting yourself with what goes on at other types of educational institutions.

If you are interested in one of the great spectrum of jobs outside of the strict realm of the mathematics research university environment (of course you will be receiving your Ph.D. from a research university), then it will, in the end, be up to you to find the openings and to make the process work in your favor. Most universities have a job interview center and an office dedicated to helping students with this process. Indeed, you may find that the Engineering School, the Business School, Arts & Sciences, and other units of the institution have their own interview centers and career offices.

7.5. The Job Interview

As previously indicated, if a school is serious about hiring you as an Assistant Professor (or sometimes even as an Instructor), then it will want to interview you. The typical scenario is this: The Chairperson phones you up and invites you for a visit. Following the suggested guidelines for dates, you yourself

---

[19]This may have to be a somewhat different letter from the one that your advisor would write for an academic position.

make the plane reservation and pay for the ticket.[20] The host department will arrange a hotel for you. And off you go.

You don't have to dress like the Queen of the May for your formal campus visit, but I would strongly urge you to think carefully about your haberdashery. It's fine when you are a student to wear turbans, dashikis, loincloths, or whatever you like, but unusual, ethnic, or offensive clothing is liable to put off a great many interviewers. Certainly T-shirts with slogans or sayings (especially vulgar, religious, or political ones) are out. T-shirts extolling the latest rock concert that you attended are out. As I have said elsewhere in this book, dress as you would to go to Thanksgiving dinner at your maiden aunt's house.

Similar comments apply to hairstyles. I appreciate that the way we wear our hair is fashion-driven. We want to fit in with our peer group and to be attractive to the opposite sex. But if you have your hair shellacked into spikes, or if you sport a green and pink Mohawk, then you are going to make a bad impression on the (unfortunately rather stuffy) people who will be interviewing you. Finally, forget body piercing. No studs in the eyebrows, no safety pins in the cheeks, no earrings (for men).[21]

You may be rapidly concluding that Professor Krantz is a tiresome old fogey; I don't blame you. I am, however, speaking here as your lawyer. If you want a job, then you must act and appear in a manner that will make your potential employer feel comfortable with you.

Of course, during the course of your visit (usually two days), you will be on display. You will be meeting faculty, administrators, the Chairperson, and perhaps a Dean or two. You'll want to be courteous, well-spoken, and thoughtful. It is a really good idea to do some research about the school before you go. Get on the internet and find out how large the school is, whether it is public or private, what type of students it has, whether it has a religious affiliation, and what happens to students after they graduate.

Think carefully (at least a week before your visit) about yourself and your attributes. You are not going to this job interview as a supplicant. You have a lot to offer. You are well-educated, you are a good mathematician, you are a solid and experienced teacher, and you bring youth, vigor, and new ideas to this department. Be prepared to describe yourself and your ideas to the people who are trying to get to know you.

---

[20]This is a nontrivial consideration. If several schools are interviewing you, then you will rack up some considerable travel bills. Of course the schools will be reimbursing you, but this may take a couple of months, so some financial planning may be in order.

[21]A friend of mine went to interview at a pretty good school. His mathematics was fine and his personality was charming, but they hated his earring, and they especially hated the fact that he was constantly talking about it—even during his formal lecture. He never got that offer.

Give some thought (in advance) to your ideas about teaching. A lot of discussion is in the air these days about teaching reform, the use of computers, group work, self-discovery, and other alternatives to lecturing.[22] Be prepared to discuss these didactic techniques and to tell what experience you have had with them. It makes a very bad impression if you respond with a glassy stare or a lot of hemming and hawing to questions that your interviewers consider essential and vital.

Think carefully about the talk(s) that you will be giving. You want to impress people with your insight and erudition, but you don't want to snow them or bore them. If you are giving a lecture at a Liberal Arts college or a comprehensive university, then it is probably inappropriate to give a fancy talk about the cohomology theory of coherent analytic sheaves. If you are giving a job talk at Harvard, then it is a big mistake to give an hour disquisition on a new method for teaching the chain rule. You must tailor your presentation to your audience. It is OK to ask your host what kind of talk they are looking for and at what level. Some schools will want you to give a sample lesson. Others will want you to lecture on a problem from the *Monthly*. The only way to find this out is to ask.

Of course I don't need to tell you that you must be on top of your mathematics. The people interviewing you want to know what kind of mathematician you are. Be prepared to discuss your thesis and what your current research interests are. If you can get involved in a serious mathematical "chalk talk" with people—standing at the blackboard and *doing* mathematics—then you are sure to make a good impression.

During your job interview visit, you want to be confident, polished, and attractive, but you also want to be humble. While you are indeed a talented individual who will be a valuable addition to the faculty, you are also a good colleague and a cooperative scholar who will fit in and make important contributions to the department. At least that is the impression that you want to give.

7.6. The Life of An Assistant Professor

In spite of some indications to the contrary, this has been primarily a book about how to become an academic mathematician. With this thought in mind, we now discuss what it is like to be an Assistant Professor and what your experiences and expectations will be.

I am assuming that you have written your thesis, received your Ph.D., spent perhaps two or three years at some good department as an Instructor

---

[22]The book [KRA1] can acquaint you with many of these teaching ideas and contains references for further reading.

or Postdoc, and have now landed an Assistant Professorship. What happens next?

The first part of the drill is the same as what you have heard before. When you accept the job (usually in conversation or *e*-mail with the Department Chairperson, but followed by a formal letter of acceptance), ask the Chairman when you need to show up for work. The answer will usually be "a few days before classes begin." You will be given an office and also a teaching assignment. And there you sit.

You will certainly want to go around and introduce yourself to the (senior) faculty who are in your field. Tell them who you are, where you got your Ph.D., who your thesis advisor was, and what you are working on (they already know most of these things, but this is how you break the ice). Make a point of hanging out in the coffee room and going to tea so that you meet people. Join groups for lunch. If faculty meetings are called, be sure to attend. Sign up for, and attend religiously, all of the seminars in your areas of interest.

But now let's examine the bigger picture. You are going to be looking at the tenure decision three or four years down the line. Your entire professional life has been a preparation for that moment. What do you do in the interim?

Let me review the three criteria for tenure—at least at a research university:

Excellent teaching: In the old days, it was sufficient for you to have a body temperature above 95° and for your teaching not to generate too many complaints. Today, at most schools, your teaching must be notably (and provably) outstanding. If it is not already so, then find out what resources are available at your new institution to help instructors with their teaching (the use of the Teaching Center, for example, is discussed in Section 3.12). Find a teaching mentor among the faculty. Have a look at the book [KRA1].

Departmental Service: In departments where research is the thing, there is a commonly held wisdom that Assistant Professors should not be burdened with too much service; they should be concentrating on developing their research program. In other departments this may not be the case. In any event, if you are called upon to serve on a committee, then you should agree to do so. Participate willingly and enthusiastically. Offer your opinion when called upon, but don't be pushy or obnoxious. I can promise you that this is one of the main ways that department members will get a sense of you. When it comes time for the tenure vote, they will remember what it was like serving on committees with you.

7.6. The Life of An Assistant Professor

Research: If you are now in a research department, then research is the main thing. [Even if you are not at a research department, your research or scholarly profile will play a significant role in your tenure decision. This will vary from institution to institution. Some colleges will be happy if you attend some conferences and give a few talks. Others want you to publish a few papers. You should find out, and find out early, what is expected at your particular school.] You must have not simply a couple of published papers; in fact, you must have a recognizable *program*, showing that you have charted your own course and created your own inroads into the subject. Your program must be good enough that if the Department writes to distinguished experts at Harvard or Paris or Göttingen and asks for an assessment of your case, they will get a thoughtful answer (not, "I'm sorry, but I never heard of this guy.").

That sounds like a tall order. How do you establish such an international reputation? Well, you must publish, and in good journals.[23] *Talk* to people. Go to conferences.[24] Give talks. Share your ideas. Collaborate with people. The strategy is to let people know who you are and what you have to offer. You want the established people—around the country and around the world—to think of you as an "up-and-coming person", one whom they are happy to assess and praise. You want to be a person whose papers are read and quoted.[25]

At a research university, the Dean's Tenure Committee would like to see perhaps ten or more published papers in your tenure dossier.[26] They would like to see those papers form a coherent,

---

[23] The book [KRA2] discusses the details of writing a paper and submitting it to a journal.

[24] Please note that mathematics conferences are *not*, generally speaking, by invitation only. Most conferences are advertised in the back of the *Notices of the AMS*. If one interests you, then you should go. There is almost always extra funding to subsidize attendance by young people. Your department will have funds to help defray your expenses as well. Don't be afraid to ask for financial assistance. I have been in the profession for thirty years and I have never used my own money to go to a conference.

[25] And now I must sound a note of caution. Assistant Professors at research universities should not write books. You may think that book authoring is scholarly work that will count positively at tenure time, but such an expectation is incorrect. In fact the people assessing your case may conclude that you are dodging the job of doing research and are writing books instead; they may be right. Writing books may be fine at four-year teaching colleges or comprehensive universities, but be sure (by asking those in authority) before you proceed with such a large undertaking.

[26] This number is going to vary quite a lot, depending on the institution—and also on the quality of the papers. As Chairperson, I have secured tenure for people in my own department who had as few as five or six papers, but it was clear that the Dean would have preferred to see more. A few individuals have received tenured Professorships at Princeton with only a couple of papers—but they were quite extraordinary pieces of work! Some Liberal Arts colleges are only looking for three or four good papers; some state universities are looking for twelve or more papers. People in applied math and statistics tend to publish more. People in low dimensional topology

focused whole, showing that you have had an impact on your subject.[27] They would like to see *at least six* letters of recommendation from ranking experts in your field (this is aside from letters about your teaching and your *gestalt* as a department member). It will help you, as you focus your energies on the tenure process, to know that this is what you are aiming for.

Grant Support: Many research universities attach considerable significance to grant support. Such support is, after all, an indication of outside recognition of one's work. Many university administrators are chemists (just because chemists tend to be "organization men"), and the subject of chemistry cannot run without a great flow of grant money. So chemists take grants for granted and they assume that you will have grants too. At my own university, it is difficult to get tenure for a young mathematician who does not have an NSF (National Science Foundation) or NSA (National Security Agency) or DOE (Department of Energy) or other research grant. Whereas chemists have dozens and dozens of places to which they can turn for funds, the pure mathematician has few. The NSF, NSA, and DOE are the most likely places, but there are a few other foundations to which one can apply. An applied mathematician can apply to DARPA, the Navy, DOE, and many other non-mathematical agencies that use high-level mathematics.

Other Forms of Outside Recognition: The Dean, and the people on the Dean's Tenure and Promotion Committee, like to see external, objective forms of recognition for a candidate's work. This is why outside letters are such an important part of the tenure dossier. Grants count a lot. And of course papers published in refereed journals are another significant yardstick. In addition, there are various awards that a young mathematician may garner. These include the Sloan Foundation Fellowship, the NSF Career Award,

---

and algebraic geometry tend to publish less. Be sure to talk to people at your school to get a clear idea of what is expected for the success of your tenure dossier.

[27]There is a delicate point that ought to be made here. Many of us grew up with the idea that it is a good thing to be a broad scholar with many interests. Unfortunately, having a catholic and wide-ranging collection of pursuits will not serve you well when it comes time for your tenure review. If you have written twelve papers—four in each of three distinct areas—then no one person can review all of your work. Each letter-writer will say, "I can comment on these papers but I cannot comment on those." That does not make a very good impression. Of course there will be exceptions to what I am about to say, but generally speaking, it is best for an Assistant Professor at a good research university to focus research efforts fairly narrowly and to burrow *very deeply*. The best of all possible situations is if all of your letter-writers say, "This candidate is the foremost expert in the world on *this subject area*, and his/her work has been the most significant in the past five years." It is fine when you are an Associate Professor, and especially when you are a full Professor, to do many different things, but when you are an Assistant Professor you should specialize.

7.6. The Life of An Assistant Professor

the Humboldt Foundation Fellowship, and the NSF Postdoc. Any of these is a feather in the cap of a young mathematician and will contribute decisively to the strength of a tenure case.

One thing that is on your side is this: If you have landed an Assistant Professorship—after a two- or three-year Instructorship—then the hiring process has already consisted of a careful scrutiny of your academic record. Your research credentials have been closely assessed. You have already been put through a "mini-tenure-review". So your department has hired you with the expectation of giving you tenure. You still must earn it. I have sketched here how you go about that task.

Of course everything I say must be filtered through the sensibilities of the institution at which you actually work. A Liberal Arts college or a comprehensive university has teaching as its main mission; its value system will be rather different from that of a research university. First of all, such an institution will certainly place greater emphasis on your teaching and departmental service. Many will want you to publish; others will be content if you are a dynamic teacher and tireless worker on behalf of the department. At a school where research plays a low-key role, the notion of writing to Harvard and Paris for letters of recommendation is out of the question. Tenure is more of an "in-house" affair at such an institution. It makes sense for you to meet early on with the Chairperson of your department to get his/her reading of what makes for a good tenure case. After all, s/he is probably going to be the one to present the case to the Dean.

Get a senior faculty mentor early on and consult with that person regularly. Your faculty mentor will play a pivotal role in your tenure decision and in educating and guiding the other faculty through the process, so you may as well work with that mentor throughout the probationary period and contribute to that education.

I have already indicated that your tenure dossier will contain six (or more) letters assessing your case. These will, for the most part, be so-called "outside letters" (from faculty at other universities, not your own) and they will be discussing your research (how could such people know anything about your departmental service or your teaching?). How are these letters obtained? Usually your Department Chairperson will sit down with you and ask you for a list of people whom you would like to have write on your behalf.[28] The Department will then generate its own list. Then the Chairperson, in consultation with some of the tenured faculty in your field,

---

[28]The Chairperson may also ask for names of people from whom you would *not* want a letter, because no matter how much of a straight arrow or good guy you are, it is always possible that you have sparked a jealousy or just plain irritated someone. Maybe someone who could be an obvious letter-writer for you is in fact an arch-enemy of your thesis advisor. You certainly don't want a letter in your tenure dossier from someone who has a grudge.

will put together a final list of potential letter-writers—drawing both from the list you submitted and from the list that the department cooked up. Depending on the university, either the Chairperson or the Dean will then write to these people and ask them to write a letter about you and your abilities. It will be made very clear that this is a letter for a tenure case, and quite specific instructions will be provided. The letter-writer will be asked to *recommend explicitly for or against tenure.* Of course you should not, indeed you should not even consider, contacting any of these people yourself. This is a process that the university handles.[29]

In summary, the tenure process is the make-or-break cornerstone of your academic career. Don't treat tenure as a black box over which you have no control. This is, after all, your life. The more you know about it, the more you help to shape the process, the better off you will be.

One very positive addition to life in the past several years has been the inception and flourishing of Project NExT, sponsored by the ExxonMobil Foundation. Project NExT is overseen and administered by the Mathematical Association of America. This is a loosely-knit organization of junior faculty across the country who want to share common interests and concerns. They are mentored by a broad cross-section of senior mathematicians who make themselves available for consultation or for just chewing the rag. The Project NExT people have their main meeting each year at the Summer MathFest (sponsored by the MAA) and reconvene, at a smaller event, during the January AMS/MAA meetings. They also organize other special events. Project NExT endeavors to inform its members about publishing, about tenure, about teaching, and about getting along in a math department.[30] It has done a lot of good for a lot of people and I encourage you to get involved—the web site is http://archives.math.utk.edu/projnext/.

In the same vein, you should develop the habit of consulting the internet newsletter called *Concerns of Young Mathematicians (CYM).*[31] Written *by* young mathematicians *for* young mathematicians, this periodical provides a forum for beginning mathematicians to exchange ideas and experiences. There are "guest articles" by more senior mathematicians (such as myself) and many articles by junior people about the job interview process, about tenure experiences, about departmental politics, and about many other matters of concern.

---

[29]It is not at all uncommon for a senior person in your own department, perhaps your mentor, to contribute a letter to your tenure dossier. Your Chairperson might ask the senior mentor to write such a letter or, under the right circumstances, you could do it. Best would be to work through the Chairperson to get such a letter if it seems appropriate.

[30]The young mathematician's home department is required to be a part of Project NExT. In particular, it is the home department that pays for travel to the NExT meetings.

[31]Check out the web site http://www.youngmath.org/newsletters.html. See also the volume [BEC].

7.7. The Tenure Clock

The American Association of University Professors mandates that a junior faculty member be considered for tenure not later than the seventh year from the Ph.D. Put a bit differently, the rule is that if an Assistant Professor is employed full-time and without interruption for seven years, then s/he is automatically tenured. Universities monitor this situation very carefully, for they only want to tenure those whom they actually *choose* to tenure.

Thus has come about the notion of the "tenure clock". This is simply the measure of where you are on that seven-year timeline. When you accept a tenure-track job at a college or university, then part of your negotiation with the Chairperson concerns the setting of your tenure clock. Of course if you are a fresh Ph.D., then your tenure clock is set at zero. If you have had a Postdoctoral position for a couple of years, then you can expect to get some credit for that time served and your tenure clock will probably be set at two years. There are a number of schools today, however, that set everyone's tenure clock to zero—regardless of what prior experience the individual may have had. According to state law in Georgia, *nobody* (not even a senior academic like myself) will get tenure immediately upon accepting a job. Everyone in Georgia must wait three years and then go through a tenure review. Institutions in the State of Florida often make even senior people wait for tenure when they take a job in the sunbelt.

In the late 1970's and 1980's and, to some extent, in the 1990's, there was tremendous competition for the best young mathematicians. Some exceptionally capable individuals would get a three-year Instructorship straight out of graduate school and then immediately get a tenured job at a good university. There were cases (only a few) of people being tenured—and even given full Professorships—straight out of graduate school. The whole notion of "tenure clock" changed drastically during this period. Now, as we all know, the economy is sluggish and most universities are doing considerable belt tightening. They can afford to be extra cautious about tenure and they are doing so. These days, most everyone waits about six years to be considered for tenure, then they wait four to six years or more to be considered for promotion to full Professor (see Section 8.5 for a discussion of this process).

7.8. What Will Be My Teaching Load?

Teaching loads vary quite a lot. Here we break the picture down by the type of school.

Perhaps we can begin with a brief historical comment. In the 1930's and 1940's, everyone—even André Weil and Oscar Zariski—taught quite a lot, but the universities were a low-key enterprise in those days. Only a small segment of the population went to college, math departments were fairly small, and those who had academic jobs considered themselves lucky to have any job at all. In addition, classes were quite small. The typical full Professor's salary even in 1950 was about one-fortieth of what I make today. So a three- or four-course teaching load together with a modest salary was not at all unusual; a five-course load was fairly common.

Beginning with the Sputnik era[32] (1957 and later) and the massive explosion of American universities in the early 1960's, the competition for top mathematicians became fierce. Universities started dreaming up new perks and incentives to attract and keep the best scholars, so salaries went up (see Section 7.10 for more information about salaries for mathematicians) and teaching loads went down. Of course there have been many shifts and cycles in the American economy in the past forty years, but salaries for academic mathematicians have remained fairly robust and teaching loads have remained fairly low. Many public universities are under pressure to raise teaching loads and some are now doing it. The information we report here is accurate as of 2003.

Junior Colleges: The typical teaching load at a junior college is five courses per semester; some have a load of six. There also may be duties in an instructional computer lab or a tutoring center.

Large State Universities: This is going to depend quite a lot on the state. In some states, the universities are quite directly answerable to the state legislature and to the voters. Since neither legislators nor voters understand much about mathematical research, they want to see some teaching, so the teaching load may be three courses per semester.

At many of the large state universities—Michigan, Wisconsin, the University of California, Penn State, Ohio State, etc.[33]—the teaching load is about two courses per term or semester. The understanding, of course, is that a significant segment of the professor's time is dedicated to departmental administration and curriculum, and of course a significant portion is dedicated to research and scholarship.

---

[32]In October 1957 the Soviet Union sent the first (unmanned) satellite, called Sputnik, around the world. This event caught the United States completely by surprise and we concluded that we were way behind in the space race. A huge initiative was immediately conceived to step up education and development in science and engineering. Thus arose the so-called "Sputnik era". It lasted about ten years.

[33]These are the state universities that think of themselves as research institutions.

Four-Year Teaching Colleges, Liberal Arts Colleges: As described elsewhere, teaching is the thing at these schools. Students pay upwards of $25,000 per year in tuition to be able to take small classes from excellent scholars, so the teaching load will be higher here than at a research institution. Three courses per term or semester is typical.

Comprehensive Universities: In the old days, these were known as "normal schools". They were institutions dedicated to teacher preparation. For a variety of societal and other reasons, there are few remaining institutions that see themselves strictly as teacher-training colleges, but the schools are still there, they still play a major role in the educational enterprise, and they employ many Ph.D. mathematicians. Such an institution has little expectation of scholarship on the part of its faculty and the teaching load is correspondingly high. Three 4-hour courses per semester or four 3-hour courses per semester is a quite common faculty duty. At some schools it will be higher.

Elite Private Universities: Of course these are the plum schools and everything about these institutions is special. A total teaching load of nine credit hours per year is typical.[34] That means three courses: two in one semester and one in the other. Faculty will commonly say that "My teaching load is 2-and-1." Some professors, like the endowed Chairs, will have a special deal and only teach one course per semester.

Some elite schools, such as Princeton University, pride themselves on being institutions where the professors actually teach. So the teaching load—even for endowed Chair Professors—is two courses per semester (usually one upper division undergraduate course and one advanced graduate course). The university compensates this hard work in other ways—by being quite liberal with sabbatical leaves (see Section 7.9), for example.

7.9. When Do I Get a Sabbatical?

The word "sabbatical" has both Greek and Hebrew antecedents. It was a custom among ancient Jewish farmers to take every seventh year off. Thus it has come about that (at least in theory) every seventh year the Professor gets a year off.

---

[34]Fields Medalist Charlie Fefferman once told me the story of a businessman and an academic chatting at a cocktail party. The businessman asked the professor how much he taught. "Nine hours," was the reply. "Well," said the businessman. "That's a long day, but it's easy work."

The purpose of the sabbatical is to give the professor some time to recharge his batteries, to learn new things, to write a book, or to conduct a collaboration. It is a way to help keep his/her scholarly program alive.

Well, like all good things in this life, the interpretation of what "sabbatical" means has evolved and it is going to vary considerably from school to school.

At the University of California, the situation is quite structured. For each six terms that you teach, you get a term off. Period. You don't have to compete for it or justify it; it's yours. You are given full salary during the time off and you can spend it building a house if you like (in fact I know one professor at a U.C. campus who did just this and he gave a student an Independent Study Course to help him).

At my own institution, after you give full service for six years, then you are entitled to a sabbatical. You can have one semester at full pay or a full year at half pay. You don't have to compete for this largesse; it is your right.

At a great many schools, especially state schools, you have to compete for your sabbatical. The Dean only has so many to give out and the number is considerably fewer than the usual "rule of seven" would dictate. As a result, you have to fill out some paperwork in order to convince the Dean that you deserve a sabbatical. This may include reminding the Dean of all the university service you have been performing. It will certainly entail making a convincing case that you are going off for a year or a semester to do something worthwhile. For example, when I was at Penn State, it was a surefire hit with the Dean if you told him you had arranged to spend a year at Harvard to sit at the feet of some Nobel Laureate to learn the latest ideas in string theory (or pick your favorite hot area of scholarly activity); if instead you told the Dean that you wanted to spend the year at the University of Southeast Idaho to write a calculus textbook with your former Ph.D. student, then he was less likely to be moved.

Let me stress that sabbaticals are not available to adjuncts and Instructors/Postdocs. The sabbatical is, along with tenure, one of the great perks of being a tenure-track academic. You don't need tenure (i.e., you don't have to be an Associate or full Professor) to take a sabbatical; you just have to have punched the time clock as a tenure-track faculty member for the right amount of time.[35] The sabbatical rules at your institution are public knowledge; be sure to apprise yourself of the law of the land and to avail yourself of this privilege whenever you can.

---

[35]Some Deans, like my own, are willing to be quite generous with sabbaticals for Assistant Professors. Of course if an Assistant Professor waits until the seventh year to apply for a sabbatical, then s/he will already be tenured (or not). Our Dean (and some other Deans) feels that a young scholar deserves this time without duties to develop a strong research program, so he is willing to give a sabbatical to an Assistant Professor with only three or four years in the saddle.

A final note is this. Faculty also have opportunities to take "leaves of absence". How does a leave differ from a sabbatical? Usually you take a leave when some other institution offers to pay your full salary and benefits for the year or the semester. This makes it easy for your Dean to let you go away for a time—because it doesn't cost the Dean anything. The Dean will start to get nervous if you ask for a second year's leave (because he will worry that you are not coming back). Most Deans will absolutely forbid a third consecutive year of leave.

You can imagine that there are certain faculty who have manifold opportunities to take leaves: If you are a world-class scholar or the recent winner of the Fields Medal or the Wolf Prize, then of course every institution would like to have you around for a while. If you do applied mathematics, then perhaps NASA or some other government agency would like to have you as a paid consultant. Maybe some big company—like Hughes Aircraft—would like to put you on staff for a year. It is easy to see how faculty may find these opportunities attractive and it is also easy to see why the Dean wants to keep them under control.

7.10. What Kind of Money Can I Make as a Professor?

Nobody ever got rich being a professor. More generally, nobody ever got rich working for someone else. Academic salaries cover a wide spectrum. I earn a pretty good salary, but there are full Professors who make half of what I earn and there are full Professors who make double what I earn. The American Mathematical Society, in its publication the *Notices*, publishes a yearly survey of professorial salaries, broken down by public and private universities and by the quality of the institutions (Group I, Group II, etc.). If you want the chapter and verse on this matter, then that is the place to look. In this section I will give you only a rough idea of what academics make. These figures are accurate as of 2003.

Adjunct Instructors: As indicated elsewhere, an adjunct is (usually) paid *by the course*.[36] There are no perks or benefits (health insurance, etc.). The typical per-course stipend can range from $2,000 to a high of $6,000. In some instances, when the instructor is somebody with special skills and the course is for a medical school or an engineering school, the fee could be as high as $10,000, but $2,000 to $4,000 is more typical.

---

[36]Some institutions, especially large state schools like Penn State, hire a considerable number of adjuncts on one-year contracts. These folks have a full teaching load and a yearly salary of about $30,000–$35,000 per year. They may have benefits as well.

Research Instructors: You can look in the *Notices of the American Mathematical Society*, *EIMS* (the publication *Employment Information in the Mathematical Sciences*), or the American Mathematical Society online job resource and see for yourself that Instructorships pay in the $40,000 to $45,000 range. At some schools, especially in a place like New York City, it could go as high as $55,000. In certain special fields, like statistics and finance (which happen to be hot and which have to compete with industry), salaries could go higher still. Applied mathematicians are often paid more than the norm.

Assistant Professors: At a research university, Assistant Professors begin in the $50,000 to $55,000 range. The salary will be lower at a four-year teaching college or comprehensive university. Of course, as noted in the **Research Instructors** paragraph, Assistant Professors in finance and statistics are paid more. Faculty in applied mathematics are often paid more.

Associate Professors: At a research university, Associate Professors begin in the $60,000 to $65,000 range. The salary will be lower at a four-year teaching college or comprehensive university. As above, Associate Professors in finance and statistics are paid more. Faculty in applied mathematics are often paid more as well.

Professors: At a research university, Professors usually begin in the $70,000 to $75,000 range, but full Professor salaries will vary widely according to the location of the university (i.e. expensive urban area or more affordable rural area) and also according to whether the institution is private or public. For people with twenty years in the business (at a research university), $100,000 can be a typical annual salary, but there are good universities today where the average professor's salary is in the low $70,000s (the University of Washington is one example) and there are also good universities where the average professor's salary is in the $140,000s (Harvard is one example).

For an endowed Chair Professorship[37] at an elite private university, a salary of $150,000 or more can be typical. There are a handful of math faculty in this country who make over $200,000 per annum. An old joke

---

[37]An endowed Chair Professorship is a special honor that is bestowed on particular faculty after an elaborate and competitive selection process. There are many more of these Chairs at private universities than at public ones. An endowed Chair usually pays the professor's salary and perks from a dedicated endowment (i.e., invested funds) and that salary is usually an especially nice one. Endowed Chair positions often come with special funds for travel or other scholarly activities. Often an endowed Chair position has a name attached to it (indicative of the source of the money). For example, Guido L. Weiss occupies the Eleanor Anheuser Chair here at Washington University (from Anheuser-Busch money). He likes to refer to himself as the "Beer Professor".

7.10. What Kind of Money Can I Make?

among seasoned faculty is that when you want to speak at a faculty meeting or other academic gathering you should **(i)** stand up, **(ii)** state your name, **(iii)** state your salary, and **(iv)** state your opinion.

A fair number of tenure-track math faculty and Postdocs—especially at research universities—have a summer research grant that pays a stipend.[38] Many faculty with special positions (Chair Professorships and the like) have special travel funds, research funds, and other "slush funds" that make life comfortable.

All of my remarks above about academic salaries are premised on the supposition that the salary is computed on a nine-month basis. This means that, contractually, the individual works for the university for nine months out of the year. For the other three months, the individual is pretty much on his/her own. The faculty member could take another job, teach summer school, get a research grant, write a book, consult, or engage in a number of other special activities.

Some schools make it possible for individuals to have twelve-month contracts. This means that the individual works for the university for twelve months out of the year and in many instances is paid 4/3 of a nine-month salary.[39] The advantage of a twelve-month contract is that you can make (considerably) more money; the disadvantage is that you have to dance to the university's tune all year long.

As you think about salary, you should bear in mind that academic life gives you a great deal of freedom. You can write books. You can consult. You can do reviewing for publishers. You have large chunks of time for yourself and you can travel or write or think (or all three!). All of these have the potential for making money, and they are all stimulating and interesting and part of the mathematical life. I know a number of mathematicians who make more money doing these sorts of "outside activities" than they do through their academic salaries, but note also that the faculty member's latitude to engage in outside activities will vary from institution to institution.[40]

---

[38] In fact roughly 30% of research-active mathematicians have research grants.

[39] Please understand that those on a nine-month contract can arrange to receive twelve paychecks per year. That is what I do, but I am still on a nine-month contract.

[40] For example, at Texas A&M University, "inventions, innovations, discoveries and improvements made with the use of System facilities or during the course of regularly assigned duties of the faculty and staff shall become the property of the System." The university and the employee split the profits. If the employee claims that s/he made the discovery during his/her free time, then negotiations with the university, most likely involving lawyers, will be involved. At most schools, faculty retain their copyrights and book royalties without interference, but if you write a best-selling calculus book and your school adopts it, then you are expected to set up a scholarship fund with the profits *from your school*.

The salary of any particular professor at any particular university is supposed to be confidential. At state universities, it is usually *not*. At any campus of Southern Illinois University, you can walk into the library, go up to the Circulation Desk, put out your hand, and say, "Gimme the book." They will hand you a binder that contains the salary of *every state employee*—even the governor. For Penn State University faculty salaries, you can just drive to Harrisburg, go to the appropriate state office, and demand to see the list. The University of California system (comprising nine campuses) has a rather rigid "step system" (i.e., a range of levels within each faculty rank). There are (as of this writing) six steps of Assistant Professor, five steps of Associate Professor, and nine steps of full Professor. It is not difficult to find out the rank and step of any of your colleagues; then you can just go on the internet and look up the corresponding salary.

At private universities, faculty salaries really are supposed to be a secret, but every year, the student newspaper at Princeton University publishes the top ten faculty salaries—and it *names names*. Don't ask me where they get the information, but it seems to be available.

A professor at a university can live a comfortable life. Such a person will never own an ocean-going yacht or a 10-room condo in Paris. As I have said elsewhere in this book, people with mathematical skills have many choices. If you want to make the big bucks, you should probably start your own company or perhaps be a consultant (or both). If, instead, you want to live a peaceful, dignified, and comfortable life, then a professorship may be for you.

Chapter 8

Afterthoughts

8.1. Research vs. Teaching

At some point in your career—and this could be in the middle of graduate school or in the middle of your probationary period as an Assistant Professor or in the middle of your tenured Professorship—you might look in the mirror and say, "I'm tired of the research life. It is effete, it is self-involved, and it is boring. I don't like the competition and I don't like the players. What is really important is teaching. With teaching I do something that has an immediate impact on society and that really helps people. For the remainder of my career, I am going to concentrate all of my effort on teaching."

Fine. The world needs more dedicated teachers. And there are over 50,000 mathematics papers published every year. If your heart is no longer in it, then your one or two papers will hardly be missed. But follow the dictum of Socrates: know yourself. Be sure you know and understand your motivations for making this shift of emphasis and effort. Are you quitting research because you've hit a roadblock, or because you've lost interest, or because you no longer have any drive or motivation? Is it because you are getting a divorce, or recovering from alcoholism, or your father has died?

All of these are strong and valid personal reasons and one of them may be the right one for you, but you have invested a lot of time, effort, and passion into developing the intellectual tools to do research and you have derived a lot from the process. You can be a great teacher and a great researcher at the same time—and this is really the best of all possible worlds, for the two activities cross-fertilize and reinforce each other in a variety of delightful and fulfilling ways.

My advice would be *not* to make any abrupt decisions. You could de-emphasize your scholarly work for a while in order to write a textbook. Or you could set aside some time to develop a new upper division curriculum in dynamical systems. But don't just *quit doing research*. That is probably not what is best for you and it is probably not what you really want.

8.2. How Do I Keep My Research Program Alive?

This is the big question. In academic life, it is easy to fall into the trap of spending the nine-month academic year completely absorbed in teaching and departmental service activities, rationalizing that the summer will be set aside for research. This is like saying—if you were in business—that you must spend 11.5 months per year fanatically pursuing the avaricious career path to success and you've set aside the two-week vacation period as a time to be nice to your family.

The sort of person described at the start of the last paragraph is, quite frankly, most likely to spend the summer preparing classes for the following year. In short, such a person is never going to do much research. S/he will never get around to it and will spend time wondering how others manage to get any research done. The great actor Humphrey Bogart was once asked how he landed so many good roles. He replied, "I just keep working, and the roles come along."

And that is the point. Of course, if you want to be a successful mathematician, it helps to be brilliant and inspired and to hang out at one of the great world mathematical centers, but you also must have the right personal habits and the right discipline. The main key to success is to have mathematics rumbling around in your head all of the time. *Always* have a problem that you are thinking about. *Always* have a project in the works. *Always* be writing something. One of my mentors used to say, "If you have something new, then write it up. If not, then write up something old. If you haven't got anything old, then read."

After you are tenured, you have considerable control over your life. If you want to be the sort of person described in the first paragraph, to dedicate yourself to teaching and to treat research as a hobby, then that is a good and rewarding career choice. God knows that our profession could use more excellent teachers. But when you are pursuing the holy grail of tenure, you are going to have to find a balance between teaching and research. *You must maintain the momentum of your research program.* That is why it is important to go to seminars and to engage other mathematicians in fruitful discourse. It is particularly helpful to have an energetic collaborator to help keep you on track and focused on your work. I have several collaborators all over the world. They are all younger than I, and almost certainly more energetic. They help to keep me alive (mathematically speaking) and they also give me a good excuse to travel. It is a good life, and a rewarding and vigorous life, and I commend it to you. See the next section for more on collaborators.

8.3. Collaborators

I have written more than 125 scholarly papers and well more than half of them have been collaborative. I think that collaboration is great. Doing mathematics is a lonely and trying endeavor. It helps tremendously to have a collaborator off of whom you can bounce ideas. A collaborator can give you courage when the going gets rough, can generate techniques and examples when you need them, and will help to keep your interest piqued. A good collaborator will be constantly generating ideas and also responding to yours. It is stimulating and exciting to have a collaborator and this gives you an excuse to travel to an exotic locale (namely, your collaborator's institution).

Today, more than half of all mathematical papers are written collaboratively. There is great joy and excitement in sharing mathematics, in discovering together, and in creating a work of permanent value alongside a collaborator and friend. I encourage you to develop collaborative relationships with other mathematicians. Just talk to people about what you are working on. If you are cooperative and sharing, then collaborations will occur naturally.

Of course a collaboration is like a marriage and you must manage it with the same delicacy. Some very fine collaborations have fallen by the wayside because of priority disputes or personal differences; this is just deplorable. Read [KRA2] and also [KRA6] to find out more about how mathematical collaborations function.

8.4. Publish or Perish

The bit of wisdom in this subsection title has been the byword of American academics for a century. If you don't publish, don't establish your scholarly reputation *in print*, or don't have a recognizable impact on your subject, then you will not get tenure and will not keep that job and retain that career that you have strived so long and so hard to achieve.

The top math departments—not just Princeton, Harvard, or Yale, but also the other top ninety-seven—will really have a hard look at your publication record at the time that your tenure dossier is reviewed. They will write to established experts in your field—all around the world—and ask for their evaluation of your case. *So your publication record and your reputation had better be such that all of the obvious people will have heard of you and be able to comment knowledgeably on your scholarly achievements.*

These days, even four-year colleges and comprehensive universities—where the emphasis is more on teaching than on traditional Germanic scholarship—will want you to have some scholarly profile. They may not require that you have reshaped a discipline, but they will want you to have a track record of some publications in refereed forums.

One of your main tasks when you first begin a tenure-track position is to find out what is expected of you in the research and publishing arena. Read the campus Tenure Document[1] and talk to the faculty—some of the recently tenured ones and some of the senior ones.

At Penn State University, there is a special room in Pattee Library called the *Penn State Room*. This room contains the professional dossier of every member of the faculty. You can go into the Penn State Room and see just what credentials Professor A possessed in order to get tenure in 1987 or what credentials Professor B possessed in order to get promoted to full Professor in 1993. Most universities do not have such a resource. You will have to go to people and *ask*. Most faculty would be willing to give you a copy of their vita. The Chairperson may maintain a book of faculty vitas and it may be available for all to see. Many faculty have their vita online. Alternatively, you can do a little detective work. Go to `MathSciNet` (the wonderful AMS online resource) and type in the name of that individual who was tenured last year. See what the publication list looked like. That will give you some idea of what you are shooting for.

8.5. Do All Assistant Professors Become Associate Professors?
Do All Associate Professors Become Full Professors?

The history of these questions is curious. There were people at UCLA and at Washington University who retired in the 1960's as Assistant Professors. There are still a good many people who retire as Associate Professors. A few words of explanation are in order.

At most research universities, getting tenure and being promoted to Associate Professor are one and the same thing. Different words are used, but it is a single process that produces both transformations at the same time. However, there are still quite a few institutions at which the processes are separate. At many of the Florida institutions, just to take an instance, an

---

[1] Every university has an official document called the "Tenure Document". It defines the concept of "tenure" for that institution and lays out how you get it and how you can lose it. Every faculty member should have some familiarity with this document. After all, it defines the parameters of one's life.

individual who is research-active is promoted to Associate Professor first—as a reward—and then tenured a couple of years later. In Georgia, by state law, nobody gets tenured before three years of university service—*nobody*. Although in past times, at certain schools, Assistant Professors were sometimes made permanent members of the faculty, this is no longer the case. Today, only Associate Professors are tenured.[2]

At most research universities, the formal criteria for promotion to full Professor are strict. The individual must have an established track record of excellent teaching, s/he must have made substantial service contributions both within the department and on university committees, and the individual *must have established and developed an important new research direction* (distinct from the research vector that got the individual tenure). In the words of my present Dean, there must (in the research component of the case) be a "smoking gun"; the lack of this component will often hang up a case of promotion to full Professor. For many scholars, the momentum provided by the thesis and by the infusion of ideas from the thesis advisor is all there is. That is enough to carry an energetic individual through to tenure. But when it comes time to do something distinctly new, to chart a new course, and establish a major individual research identity, the resources just are not there, and the promotion to full Professor simply does not happen.

Some Deans (for example, a Dean who worked at Washington University twenty years ago) will say, "We bought the barn; we may as well paint it," which means that everyone should eventually be a full Professor. The philosophy, which makes good sense, is that if an individual turns 50 or 55 and realizes that s/he is never going to be a full Professor, then the individual is likely to become discouraged and will not feel like contributing to the welfare of the department. Not promoting an individual could be like turning him/her into a nonperson, so it is best to find the right time for each individual and make that promotion happen. The opposing point of view is that a university must maintain its scholarly standards; if it promotes the wrong people then it is lowering the overall quality of the university. I would say that most—although not all—university faculty adhere to the first of these points of view.

It is an interesting fact that, about twenty years ago, Harvard University effectively eliminated the rank of Associate Professor. The philosophy was that if you are good enough to be tenured at Harvard, then you are good

---

[2]There are some famous cases of full Professors at Harvard who never earned Ph.D.'s. Harvard used to have a special system for creating "Junior Fellows". These were gifted individuals who went directly from being students at Harvard College to being faculty. They were fully vested, tenured Professors in the Harvard Math Department (or in some other department), but they never earned their Ph.D.'s and never had to fight their way up the professorial ranks.

enough to be a full Professor. There is no sense to shilly-shally around with the middling rank. I know of no other institution that has adopted this policy explicitly, but many of the elite private research universities rush people through the Associate Professor rank and into full Professor status. By contrast, many of the middle-level universities in effect *require* an individual to spend a certain number of years at the Associate Professor level—often six years or more.

The highest academic rank is full Professor. Of course, a full Professor can still strive to be an endowed Chair Professor, but the truth is that most universities have very few of those cherished positions. So, for most of us, a full Professorship is the end of the line. There are no more domains to conquer. One ultimately realizes that the be-all and end-all of academic life is to do good work that one can be proud of; in the end, the ultimate judge of the worth of that work is oneself. And that is not such a bad thing.

8.6. What If I Don't Get Tenure?

This is not good, but it is not the end of the line. This author did not get tenure at his first job, and I know people who are now eminent professors at Harvard and Princeton and other fine schools who did not get tenure in their first shot.

Clearly you are at a disadvantage if you did not get tenure at school X and seek a job at school Y. School Y will no doubt wonder what went wrong at school X (and there is really no way to hide it—academic mathematics is a fairly small world). If you are lucky, as I was, people (including your thesis advisor) will rally to your support and explain to the world that school X has made a huge error and that now many schools have a great opportunity to make a killing.

But, realistically speaking, if you do not get tenure at school X, then you will probably end up at another school of a distinctly lesser ranking, and without tenure. You will have to battle your way back up the ranks. It can be done. I've done it and others have done it, but it is tough.

Sometimes you don't get tenure because you don't deserve tenure. It's as simple as that, and you must pick yourself up and battle on. Other times the school simply makes a mistake. I know of cases where terrific people did not get tenure because nobody understood their work, or somebody thought there was an error in the work (and there was not), or somebody did not get along with his/her mentor. Some very good mathematicians have not been awarded tenure because they could not teach. Others have not secured tenure because they won a teaching award and therefore were not deemed a good bet for a long-term, productive research career. Really! All kinds of things can go wrong and it is too bad, but if you want to stay in the profession, then you can. You will have to swallow your pride and start again, but you can do it. I can say in my own case that it was well worth the struggle, but this is something you will have to decide for yourself (with the aid of your supporters and your friends).

8.7. What If Graduate School Was a Big Mistake?

University counselors tell me that the most common psychological malady among graduate students is the "impostor syndrome". The student becomes convinced that s/he is a fake, that it was only a clerical error that got him/her into graduate school in the first place, and that the student is in the process of making a huge ass of himself/herself. As soon as the chicanery is uncovered, as soon as the jig is up, this poor graduate student will be out on the street.

This condition is a form of depression. Let me assure you that it does not only affect graduate students. It affects everyone in academics. It affects me. After all, here we are in a top-notch university, surrounded by some of the best minds in the world—Nobel Laureates, MacArthur Prize holders, and other distinguished scholars of international repute. What in the world gives us the *chutzpah* to suppose that we belong here?

Graduate study—the path to a Ph.D.—is a long haul, with many a place for a misstep or an error. There is lots of room for mistakes and misconceptions. We have all been there. Going to college for four years, then graduate school for five years or more, then six years of probation as an Assistant Professor before tenure is perhaps too much for some people. After all, you could major in computer science and get an MBA in about six years total and go off to a job where you would start at $70,000 per year. Perhaps that holds some appeal for you.

On the positive side, doing scholarly mathematics and teaching is a wonderful, rewarding, and fulfilling life. You get to do what you love, you get a lot of time to pursue your own interests, and you get a certain amount of respect and admiration.

The good news for you is that, if you choose to pursue the path to become an academic mathematician, you will be surrounded by people who are in the same boat and you will be surrounded by advisors and mentors who have been through the same process. They understand your trials and tribulations and they can counsel you and help you through them.

Again, the message is to talk to people. The worst thing you can do is to let yourself become isolated. In his remarkable book [HAL], Philip Hallie argues that the way that people control and torment others is by *isolating them*. You can do this to yourself also. It is a trap and you must avoid it.

The mathematics community can be nurturing and caring. You can be the beneficiary, get your education, and live a good life, but this is a choice you must make. I hope that this book has given you a detailed sense of the process so that you can make the right choices for yourself.

8.8. What If I Only Want a Master's Degree?

Many schools *require* that you get a Master's Degree (either an M.S. or M.A.) before going on to a Ph.D. Many do not. Many of the top programs do not even have a formal Master's curriculum; they only award Master's Degrees in special circumstances (see below).

How does a Master's Degree differ from a Ph.D.? The main difference is that the former does not require you to do any original research and the latter does. For a Master's Degree, you are usually required to do a certain amount of coursework. You may be asked to take some (but probably not all) of the qualifying exams, or you may be asked to write a (nonresearch) thesis; this tract would usually be shorter than a research (Ph.D.) thesis. The Master's thesis would be expository in nature, perhaps explaining a theorem or two from some recently published paper or explaining a new trend in mathematics, such as wavelet theory or string theory.

In today's world, the Master's Degree is often the entry level of training for a job in the tech sector. At some schools you can earn an accelerated Master's in one year; two years is more typical. Some schools have a three-and-two program which gets you your Bachelor's and your Master's in a total of five years. At some schools a really accelerated undergraduate can take a few exams in the senior year and end up with two degrees (a Bachelor's and a Master's) in just four years.

It is, in fact, quite easy to collect degrees, and I have known people who made an avocation of it. Having all of these sheepskins has no credence in the real world; in the end it's just a game. If your goal is to become an academic mathematician, then the only degree that matters is the Ph.D. After you've been out in the profession for ten years, then even that degree doesn't matter much any more (except for technical reasons—some universities will not employ any faculty member without a Ph.D.).[3]

A final point is this. The Master's Degree has become a bit tarnished in the past few decades because it is often awarded as a consolation prize. If a

---

[3] I had a friend in graduate school who quit going to his high school classes when he was accepted for undergraduate school at Harvard. He quit going to his Harvard classes when he was accepted for graduate work at Princeton. And he quit doing his work at Princeton when he got a job as an Assistant Professor. So here he was, a faculty member at a good university with no high school degree, no college degree, and no graduate degree. When it came time to tenure him—and he was eminently deserving of tenure—they told him he needed a Ph.D., so he had to go back and make up all of those missing degrees.

student washes out of the Ph.D. program—either because s/he cannot pass the quals or because the thesis work runs into a dead end—then frequently it can be arranged for the student to take away a Master's Degree. Such a Master's Degree is just as good as any other Master's Degree; it is simply earned under more trying circumstances.

In today's job market, the Master's Degree can be quite attractive; it could be the perfect qualification for the career you seek. A Master's degree is often required to do adjunct teaching (i.e., teaching course-by-course for a fixed fee) or to be employed full-time while A.B.D. The Master's Degree won't get you a tenured Professorship at Harvard, but it can get you a well-paying and satisfying job as an actuary, in a high-tech firm, or working in another scientific arena.

8.9. Rounding Out the Graduate School Experience

If you choose the right graduate school and the right thesis advisor (and I hope that this book will have aided you in the process), then you will leave your Ph.D. program with a collection of lifelong friends—most notable among them your thesis advisor. And you will have an education that is your entrée to becoming a mathematician.

There is an important epistemological point here. Possessing a Ph.D. in mathematics does not *make you a mathematician*, but it gives you the right to try. The hard fact is that 90% of mathematics Ph.D.'s never go on to do any additional (original) scholarly work. This is either because they were more dependent on their thesis advisors than they should have been, and they don't have the drive to push their program any further; or perhaps because they don't have the tenacity and the courage to develop their own scholarly path. Or else they just can't make the time to do it. Of the 10% who actually do go on to do additional research, at least half of those never wander far from their thesis. Again, this could reflect lack of drive, lack of imagination, or lack of talent.

I state these facts just because they are facts. We who earned our degrees at Princeton in 1974 all thought we were future Fields Medalists and off to jobs at Harvard and MIT. In fact, one of my classmates *did* get the Fields Medal and one of us now works at MIT. But most of us are living good lives at good places. Nothing more.

This is the way the world shakes out. Your Ph.D. in mathematics will get you a good job and you make of it what you will. You may become a famous researcher, or you may become part of a research group at a big company, or you may work in a government agency, or you may become a great teacher. All important and fulfilling vocations.

One of the best things about a degree in mathematics—at *any level*—is that it opens many doors and closes few of them. It gives you a world of opportunities from which to choose. What more could one ask?

Part 5

The Elements of Mathematics

Chapter 9

The Mathematics I Need to Know

The qualifying exams define in broad strokes what you need to know to go on to the next level (i.e., the writing of the thesis). Most departments will have a detailed syllabus to go with each qualifying exam. These are readily available and you can use them as a blueprint for your studies in the first couple of years of graduate work. In the sections below I include the Washington University syllabus (a list of topics, really) for each of our qualifying exams.

There is much here that is left unsaid. Successful mathematics is usually practiced by creating a synthesis of different areas; let me assure you that no class is going to teach you how to create such a synthesis. If you are lucky, you will have a thesis advisor (as I did) who will set a great example and will point you in the right directions. But, in the end, this cross-fertilization is something that you must do for yourself.

In the present chapter we shall describe—in a sense—the different parts of a basic mathematics education. These are not the things that you need to know when you walk in the door of your new graduate school. Rather, this is the program you will be engaging in. It will prepare you for the quals and point you beyond them. Of course the level of detail in this chapter will be slight, but the neophyte reader should be able to come away with a feeling of what a graduate education is all about, what needs to be known, and the nature of the task that lies ahead.

It would be futile and infeasible for me to attempt to actually describe all of the topics in the year-long algebra qualifying exam course, the year-long complex analysis qualifying exam course, and so forth. The book [GAR] attempts to give a comprehensive description, but it only succeeds in scratching the surface. The book [GER] offers a somewhat different point of view. What we endeavor instead to accomplish is to give a notion of the flow of ideas in each subject area. The main message here is that mathematics is no longer a sequence of sound bites; instead it is a coherent whole with a great deal of structure and depth. Thus each of the subject area treatments below—real analysis, complex analysis, geometry/topology, and algebra—only gives the *first steps* of the graduate study process. This is followed by a complete list of qualifying exam topics. We hope that the union of these two modest offerings will give the student a feeling for the depth and substance of what s/he is trying to master for the qualifying exams.

Remember that you are leaving a rather desultory portion of your education (the amassing of facts and elementary techniques) and launching into a very exciting part (learning to actually *create* new mathematics). The pieces of mathematics described here are the most basic components of your toolkit. They will set you on the road to doing research. The rest is up to you.

9.1. Real Analysis

There are two main points to the qualifying exam course in real analysis. One is to learn Lebesgue measure theory and the other is to learn basic functional analysis. Let us now describe what these two general subject areas entail.

In many contexts of mathematical analysis, the nub of the problem turns out to be establishing an equality of the kind

$$\lim_{j \to \infty} \int_X f_j(x)\,dx = \int_X \lim_{j \to \infty} f_j(x)\,dx. \tag{$*$}$$

Thus commuting a limit with an integral is a matter of intense study. Certainly questions about convergence of Fourier series, regularity results for partial differential equations, existence theorems for differential equations, and many other basic analytical questions boil down to an identity like $(*)$. Different theories of the integral give rise to different means for thinking about $(*)$.

The *Riemann integral* of a function f on an interval $[a, b]$ is defined by cutting up the *domain* of f. Thus are created Riemann sums and the integral is defined to be the limit of these Riemann sums (when that limit exists). One remarkable feature of the Riemann integral is that those functions for which the Riemann integral exists can be completely characterized: A function f on an interval $[a, b]$ is Riemann integrable if and only if it is bounded and the set of its discontinuities forms a set of measure zero. [Of course one must decide what "measure zero" means, but this will become clearer in the discussion below.]

The basic convergence result for the Riemann integral is that (∗) will hold provided that $f_j \to f$ uniformly on the domain of integration. The condition of uniform convergence is rather restrictive and often does not come up naturally in practice. This weakness gave rise, historically, to a desire for a more powerful theory of the integral.

The so-called *Lebesgue integral* (Tonelli also deserves credit for its invention) takes a new approach to the construction of an integral. The basic new feature is that one divides up the *range* of the function instead of the domain. This is not the proper venue in which to present all of the details of the construction, but we can describe some of the main features.

It is a metatheorem in mathematics that it is impossible to integrate all functions. One must restrict attention to a standardized class. The so-called "measurable functions" turn out to be well-suited to our purposes. Measurable functions are a subtle generalization of continuous functions. Recall that a function $f : \mathbb{R} \to \mathbb{R}$ is continuous if and only if $f^{-1}(U)$ is open whenever U is open. Now we say that f is *measurable* if and only if $f^{-1}(U)$ belongs to a specified class of *measurable sets* whenever U is open. Here the measurable sets is a class of sets that is closed under countable union, countable intersection, and complementation and that contains all of the open sets. All sets of zero length are also included in the class of measurable sets. We know how to assign a length or measure to any open interval. Thus we can assign a measure $m(S)$ to any open set S. Now elementary reasoning by complementation and limit processes enables one to assign a measure to any measurable set.

A function is called *simple* if it is measurable and if it only takes finitely many values. If the simple function f takes value λ_j on the set A_j, then we define
$$\int f\, dx = \sum_j \lambda_j m(A_j).$$

Next, one shows that any nonnegative, measurable function f is the monotone increasing limit of simple functions s_j. And one defines
$$\int f\, dx = \lim_{j \to \infty} \int s_j\, dx.$$

Finally, one extends the definition to arbitrary measurable, real-valued f by linearity. The definition of the integral of a complex-valued f then follows easily by complex linearity.

We can indicate some of the flexibility of the Lebesgue construction with these three theorems:

Theorem [Lebesgue Monotone Convergence]: *Let $0 \leq f_1 \leq f_2 \leq \cdots$ be nonnegative, measurable functions. Then*

$$\lim_{j \to \infty} \int f_j \, dx = \int \lim_{j \to \infty} f_j \, dx \, .$$

Theorem [Lebesgue Dominated Convergence]: *Let $f_1, f_2, \ldots$ be measurable functions. Suppose that there is a nonnegative function g that is known to be integrable and such that $|f_j(x)| \leq g(x)$ for each j and every x. Then*

$$\lim_{j \to \infty} \int f_j \, dx = \int \lim_{j \to \infty} f_j \, dx \, .$$

Theorem [Fatou's Lemma]: *Let f_j be nonnegative, measurable functions. Then*

$$\int \liminf_{j \to \infty} f_j(x) \, dx \leq \liminf_{j \to \infty} \int f_j(x) \, dx \, .$$

In your qualifying exam course on real analysis you will learn about measurable sets and functions, you will learn the construction of the Lebesgue integral, and you will learn about these three theorems. Most importantly, you will learn *how to use* these three theorems. In most graduate programs, the real analysis qualifying exam is the most difficult. This is because real analysis is less about *ideas* and more about *technique*. It is essential, as you study for your real analysis qual, that you do a great many exercises that will drill you in the use of these "big three" theorems.

The other half of the real analysis qualifying course is functional analysis. The premise of functional analysis is that one gains a great deal of power and insight if one studies not one function at a time but rather entire *families* of functions.

Thus we define a *Banach space* to be a vector space equipped with a norm so that the space is complete in the induced topology. The most interesting Banach spaces are spaces of functions. For example, the space $C[0,1]$ consisting of the continuous functions on the interval $[0,1] \subseteq \mathbb{R}$ equipped with the norm

$$\|f\| = \sup_{x \in [0,1]} |f(x)|$$

9.1. Real Analysis

is a Banach space. The space $L^1(\mathbb{R})$ of Lebesgue integrable functions on the real line equipped with the norm

$$\|f\| = \int |f(x)|\, dx$$

is a Banach space. There are of course many other important examples.

The *dual* of a Banach space X is defined to be the collection of all continuous linear functionals on X. The dual is denoted X^*, and is itself a Banach space. The space X^* has considerably more structure than X and is the key to much of the depth of functional analysis.

If $\lambda \in X^*$ is an element of the dual space of the Banach space X, then we define the *norm* of λ to be

$$\|\lambda\| = \sup_{\substack{x \in X \\ \|x\| \leq 1}} |\lambda(x)|.$$

The three key results of elementary Banach space theory are these:

Theorem [Open Mapping Principle]: *Let $T : X \to Y$ be a surjective, continuous linear mapping of Banach spaces. Then T is open in the sense that $T(U)$ is an open subset of Y whenever U is an open subset of X.*

Theorem [Uniform Boundedness Principle]: *Let T_j be continuous linear functionals on the Banach space X. Then either*
  (i) *The functionals T_j are uniformly bounded in norm, i.e., $\|T_j\| \leq C$ for all j;*
  or
  (ii) *There is dense open subset $U \subseteq X$ such that*

$$\limsup_{j \to \infty} |T_j(x)| = +\infty$$

  for each $x \in U$.

Theorem [Hahn-Banach Theorem]: *Let X be a Banach space and $Y \subseteq X$ a Banach subspace. If $T : Y \to \mathbb{C}$ is a continuous linear functional then there is a linear extension $\widetilde{T}$ of T such that $\widetilde{T} : X \to \mathbb{C}$, $\widetilde{T}|_Y = T$, and $\|\widetilde{T}\| = \|T\|$.*

The proofs of these three results are fairly straightforward and their meaning is easy to apprehend. So, again, mastery of this material is a matter of *technique*. You must do a great many exercises so that you become facile with these three results, know when to apply them, and can use them effectively.

There is a special class of Banach spaces that enjoys a useful additional structure. These are the *Hilbert spaces*. A linear space H is called a Hilbert space if it is equipped with an inner product $\langle x, y \rangle$ such that the norm $\|x\| \equiv \sqrt{\langle x, x \rangle}$ makes H into a complete space.

Possession of an inner product is an important convexity condition that turns out to be very powerful. Perhaps the most significant basic result is the *Riesz representation theorem*:

> **Theorem [Riesz Representation]:** Let H be a Hilbert space and T a continuous linear functional on H. Then there is an element $t \in H$ such that
> $$Tx = \langle x, t \rangle$$
> for every $x \in H$.

The other key idea in elementary Hilbert space theory is that of an orthonormal basis. [For the sake of the present discussion, we restrict attention to separable Hilbert spaces, so that all bases are countable.] Let $\{\varphi_j\}$ be elements of the Hilbert space H. If

(a) $\|\varphi_j\| = 1$ for each j;
(b) $\langle \varphi_j, \varphi_k \rangle = 0$ whenever $j \neq k$;
(c) $x = 0$ whenever $\langle x, \varphi_j \rangle = 0$ for all j;

then we say that $\{\varphi_j\}$ is a *complete orthonormal system* (or basis) for H.

A basic result about complete orthonormal systems is this:

> **Theorem [Riesz]:** Let $\{\varphi_j\}$ be a complete orthonormal system for the Hilbert space H. If $x \in H$ then let $a_j = \langle x, \varphi_j \rangle$ for each $j = 1, 2, \ldots$. The following results hold:
>
> **(1)** If $x \in H$ then, for any M,
> $$\left\| \sum_{j=1}^{M} a_j \varphi_j \right\| \leq \|x\|;$$
>
> **(2)** We have
> $$\sum_{j=1}^{\infty} |a_j|^2 = \|x\|^2;$$
>
> **(3)** The partial sums $\sum_{j=1}^{M} a_j \varphi_j$ converge to x in the Hilbert space topology.

9.1. Real Analysis

In fact this last theorem is the basis for the theory of Fourier series in the context of square-integrable functions.

We conclude this discussion of Hilbert space with a brief treatment of the concept of projection. Let H be a Hilbert space and K a Hilbert subspace. If $x \in H$, then it is possible to write

$$x = k + k^\perp, \qquad (\star)$$

where $k \in K$ and $k^\perp$ satisfies $\langle k^\perp, y \rangle = 0$ for every element $y \in K$. The decomposition $(\star)$ is unique.

The excellent book [FOL] is the modern paradigm for what a qualifying exam course in real analysis should be. It contains all of the fundamental material that we have just described, presented in a particularly elegant and compelling manner. It also contains topics in harmonic analysis, probability theory, abstract measure theory, and integration on groups. An instructor will cover these additional topics selectively and as time permits. They may or may not appear on the qualifying exam.

A typical list of topics for the Real Analysis Qualifier is this:

- σ-algebras
- abstract measure theory
- integration
- bounded variation
- absolute continuity
- complex measures
- Radon-Nikodým theorem
- Baire category theorem
- dual spaces
- Hahn-Banach theorem
- Hilbert space
- orthogonality, projections
- L^p spaces
- distribution functions
- interpolation of L^p spaces
- the dual of $C_0(X)$
- Fourier series
- the Fourier transform
- Plancherel's theorem
- Fourier analysis on lca groups
- Sobolev spaces
- Lebesgue measure theory on $\mathbb{R}^N$
- measurable functions
- product measures, Fubini's theorem
- monotonic functions
- signed measures
- differentiation of measures
- Lebesgue's theorems
- Fatou's lemma
- Banach spaces
- uniform boundedness principle
- open mapping principle
- Riesz representation theorem
- spectral theory
- duality of L^p spaces
- weak L^p
- positive functionals on $C_c(X)$
- the Haar integral
- summability methods
- L^2 theory in Fourier analysis
- Fourier inversion
- distributions

9.2. Complex Analysis

Complex variables is the other cornerstone of the analysis portion of the qualifying exams. In this subject we study holomorphic and meromorphic functions of a complex variable.

Let $U \subseteq \mathbb{C}$ be a connected open set (usually called a *domain*). Let h be a continuously differentiable function on U. We write the complex variable z as $z = x + iy$ and thus have a natural identification of the complex numbers with $\mathbb{R}^2$. We say that a continuously differentiable function h is *holomorphic* on U if it satisfies the Cauchy-Riemann equations:

$$\frac{\partial u}{\partial x} = \frac{\partial v}{\partial y} \quad \text{and} \quad \frac{\partial u}{\partial y} = -\frac{\partial v}{\partial x}.$$

If we write

$$\frac{\partial}{\partial z} = \frac{1}{2}\left(\frac{\partial}{\partial x} - i\frac{\partial}{\partial y}\right) \quad \text{and} \quad \frac{\partial}{\partial \bar{z}} = \frac{1}{2}\left(\frac{\partial}{\partial x} + i\frac{\partial}{\partial y}\right),$$

then a continuously differentiable function h is holomorphic on U if and only if

$$\frac{\partial}{\partial \bar{z}} h \equiv 0 \quad \text{on } U.$$

It turns out that a holomorphic function is infinitely differentiable. Thus a holomorphic function h will have only $\partial^j/\partial z^j$ derivatives, and we abbreviate these derivatives by $h^{(j)}$.

The central idea in this subject is the Cauchy theory. The Cauchy theory stems from the *Cauchy integral theorem*, which is actually just a variant of Green's theorem rendered in complex notation. In what follows, we let $\overline{U}$ denote the closure of a domain U, that is, the union of U with its boundary curve γ. Now we have:

> **Theorem:** Let U be a *domain* in the plane having as its boundary a single, positively oriented, continuously differentiable curve γ. Let h be a function that is continuous on $\overline{U}$ and holomorphic on U. Then
>
> $$\oint_\gamma h(\zeta)\, d\zeta = 0. \qquad (*)$$

Here $\oint$ denotes the complex line integral, which is a complex-analytic variant of the standard line integral that you learned about in multi-variable calculus. See [GRK] for this concept and all of the ideas discussed in the present section.

One can derive immediately from $(*)$ the *Cauchy integral formula*:

9.2. Complex Analysis

Theorem: *Let U be a connected open set (a domain) in the plane having as its boundary a single continuously differentiable curve γ. Let h be a function that is continuous on $\overline{U}$ and holomorphic on U. Then, for any $z \in U$,*

$$h(z) = \frac{1}{2\pi i} \oint_\gamma \frac{h(\zeta)}{\zeta - z}\, d\zeta. \qquad (\star)$$

Formula $(\star)$ is actually the wellspring of many of the basic facts about holomorphic functions. For example, because the Cauchy kernel $1/(\zeta - z)$ is real/complex analytic, we may immediately derive the fact that a holomorphic function has a power series expansion:

$$h(z) = \sum_{j=0}^{\infty} \frac{h^{(j)}(p)}{j!}(z - p)^j$$

for z in a suitably small disc $D(p, r) \equiv \{z \in \mathbb{C} : |z - p| < r\}$ centered at $p \in U$. Here $h^{(j)}$ denotes the jth complex derivative of h.

Another result that follows from direct estimation of formula $(\star)$ is the Cauchy estimates:

Theorem: *If h is holomorphic in a neighborhood of the closed disc $\overline{D}(P, r)$ with center P and radius r then, for all $j \geq 0$,*

$$|h^{(j)}(P)| \leq \frac{M \cdot j!}{r^j}.$$

Here $M = \sup_{z \in \overline{D}(P,r)} |f(z)|$.

From this one can prove Liouville's theorem: if h is holomorphic on the entire plane and bounded, then $h \equiv 0$. For the Cauchy estimates on a large disc $D(z, r)$ imply that

$$|h^{(1)}(z)| \leq \frac{M}{r}.$$

Letting $r \to \infty$ gives that $h^{(1)}(z) = 0$. But then, since z was arbitrary, h must be constant.

And now one can derive the fundamental theorem of algebra: that any non-constant polynomial has a root. Suppose that p is a polynomial in z of degree at least 1. Seeking a contradiction, we suppose that p does not vanish. Then $h = 1/p$ is a holomorphic function on the entire plane that is bounded. It follows that h is a constant, and that is a contradiction.

Of course those are just the basics, and there are many more essential results that follow immediately from the Cauchy theory. Here are some of them:

Theorem [Weierstrass]: *Let f_j be a sequence of holomorphic functions on a domain U that converges uniformly* on compact sets. Then the limit function f is also holomorphic.

Theorem [Morera]: Let f be a continuous function on a domain $U \subseteq \mathbb{C}$. Suppose that

$$\oint_\gamma f(\zeta)\, d\zeta = 0$$

for every curve γ whose trace is a triangle that lies entirely in U. Then f is holomorphic.

Theorem [The Argument Principle]: Let f be a holomorphic function on a domain U and suppose that $\overline{D}(p,r) \subseteq U$. Assume that f does not vanish on $\partial D(p,r)$. Then

$$\frac{1}{2\pi i} \oint_{\partial D(p,r)} \frac{f'(\zeta)}{f(\zeta)}\, d\zeta = (\#\ \text{zeros of}\ f\ \text{inside}\ D(p,r))\,.$$

The argument principle has a number of useful corollaries, including Hurwitz's theorem (below) and Rouché's theorem.

Theorem [Hurwitz]: Let f_j be a sequence of nonvanishing holomorphic functions on a domain U that converges uniformly on compact sets. Then the limit function is either nonvanishing or identically zero.

The *maximum principle* says that a nonconstant holomorphic function f on a domain U cannot have the property that $|f|$ has a local maximum at any point of U. This simple idea has many profound consequences. One of these is Schwarz's lemma:

Theorem: Let f be a holomorphic function on the unit disc which is bounded by 1. Assume that $f(0) = 0$. Then $|f(\zeta)| \leq |\zeta|$ for all ζ and $|f'(0)| \leq 1$. If equality holds in the first estimate for some $\zeta \neq 0$ or if equality holds in the second estimate then f is a rotation of the disc.

This result has strongly influenced modern geometric function theory and has led to the study of curvature in the context of complex analysis (thanks to Ahlfors).

One immediate consequence of Schwarz's lemma is this useful result:

Theorem: Let $\phi : D(0,1) \to D(0,1)$ be a conformal (i.e., one-to-one, onto, holomorphic) mapping. Then f has the form

$$f(\zeta) = e^{i\tau} \cdot \frac{\zeta - \alpha}{1 - \overline{\alpha}\zeta}$$

for some $\tau \in [0, 2\pi)$ and some complex number α of modulus less than 1.

After the basic Cauchy theory, there are a number of big topics that provide real depth in the subject of complex analysis. One of the pre-eminent of these is the Riemann Mapping Theorem:

> **Theorem:** Let $U \subseteq \mathbb{C}$ be a simply connected domain which is not all of $\mathbb{C}$. Then there is a conformal mapping
>
> $$\phi : U \to D(0,1)$$
>
> which is both one-to-one and onto.

This is a startling result, and completely sharp. Koebe generalized the theorem in the following astonishing way. Let $\mathcal{R}$ be any Riemann surface (i.e., a 2-dimensional manifold that has a complex-analytic structure). Then, since $\mathcal{R}$ is a 2-manifold, it has a universal covering surface M. Now M is simply connected, and it inherits a Riemann surface structure (by local pullback) from $\mathcal{R}$. What could M be?

It turns out that there are only three possibilities: M could be (conformally equivalent to) **(i)** the disc, **(ii)** the entire plane $\mathbb{C}$, or **(iii)** the Riemann sphere. It turns out that case **(ii)** only comes up when the original Riemann surface was either the entire plane or a cylinder; also case **(iii)** only comes up when the original Riemann surface was a sphere. So Koebe's uniformization theorem says that any interesting Riemann surface is covered by the disc. In the case that the Riemann surface is simply connected, then of course the covering map becomes one-to-one and onto, that is, a conformal equivalence.

A lot can be said about the zero sets of holomorphic functions. First of all, the zeros of a holomorphic, not-constantly-zero f on a domain U cannot have any accumulation point in U. In case one has growth conditions on f, then one can say more. For example, if f is a bounded holomorphic function on the unit disc $D(0, 1)$ and if $\{p_j\}$ are its zeros (counting multiplicity), then

$$\sum_j (1 - |p_j|) < \infty.$$

There is a converse to this statement as well.

In fact, one can sharpen the first result of the last paragraph. If U is a domain and $\{p_j\}$ is any discrete set in U—that is to say, $\{p_j\}$ has no interior accumulation point—then (thanks to Weierstrass) there is a holomorphic function on U that vanishes at the p_j and nowhere else. There are various sharpenings of this result due to Mittag-Leffler.

Considerable time may be spent studying meromorphic functions—these are functions which are locally the quotients of holomorphic functions. Thus a meromorphic function can have a singularity, called a *pole*. A compact Riemann surface cannot, of course, support any nonconstant holomorphic functions, but it does possess a large collection of meromorphic functions—enough to tell us a great deal about the topology of the surface (see, for example, the Riemann-Roch theorem).

A lovely and powerful way to round out a first-year graduate course in complex variables is with the prime number theorem. Long-sought since the time of Gauss, this result was finally established independently by de la Vallée Poussin and Hadamard (using techniques of complex analysis!) in 1896.

Theorem: Let $\pi(n)$ represent the number of prime integers less than or equal to the positive integer n. Then
$$\lim_{n \to \infty} \frac{\pi(n)}{n/\log n} = 1 \,.$$

A typical list of topics for the Complex Analysis Qualifier is this:

- complex numbers
- harmonic functions
- $\frac{\partial}{\partial z}$ and $\frac{\partial}{\partial \bar{z}}$
- elementary functions ($e^z, \log z, \arg z$)
- power series
- Liouville's theorem
- Schwarz-Pick lemma
- Laurent series
- Casorati-Weierstrass theorem
- Cauchy integral theorem
- Runge's theorem
- existence of primitives and logs
- real integrals by residues
- argument principle
- local mapping theorem
- Ascoli-Arzéla theorem
- Riemann mapping theorem
- Schwarz-Christoffel formula
- theta functions
- Weierstrass factorization

- complex differentiation
- harmonic conjugates
- linear fractional transformations
- elementary Riemann surfaces
- complex integration
- maximum modulus principle
- mean value property (and converse)
- classification of singularities and zeros
- residues
- Cauchy integral formula
- simply connected domains
- residue theorem
- Poisson integral formula
- Rouché's theorem
- normal families
- Montel's theorem
- Schwarz reflection principle
- mappings of rectangles, elliptic integrals

- Mittag-Leffler theorem
- Blaschke products
- factoring by Blaschke products
- Hadamard theorem on lacunarity
- product representation
- Stirling's formula
- Riemann ζ-function
- functional equation for ζ
- periodic functions
- covering spaces
- fundamental domain of modular group
- σ-function
- congruence subgroup mod 2
- analytic continuation
- Picard's little theorem
- subharmonic functions
- growth and zeros of entire functions
- univalent functions
- infinite products
- Weierstrass theorem
- Jensen's formula
- boundary theory of H^p functions
- inner and outer functions
- gamma function
- Bohr-Mollerup theorem
- beta function
- Euler product formula
- prime number theorem
- elliptic functions
- modular group
- Weierstrass $\mathcal{P}$-function
- modular function
- fundamental domain for congruence subgroup
- monodromy theorem
- Picard's great theorem
- solution of Dirichlet problem
- Nevanlinna characteristic

9.3. Geometry/Topology

Topology is concerned with homeomorphisms—maps which preserve open sets. Topology does not recognize rigid structures. Modern differential geometry is concerned with metric rigidity and the invariants that it preserves. Chief among these is curvature.

Topology

A *topology* on a set X is a collection $\mathcal{U}$ of subsets that is closed under finite intersections and arbitrary unions and such that $\emptyset$ and X are elements of $\mathcal{U}$. The elements of $\mathcal{U}$ are called *open sets*. The complement of an open set is called a *closed set*. If $x \in X$ and $U \in \mathcal{U}$ is an open set such that $x \in U$ then we say that U is a *neighborhood* of x. We will often denote the topological space by the ordered pair $(X, \mathcal{U})$.

Examples of topologies are these:

(1) Let $X = \mathbb{R}$ and $\mathcal{U}$ be the collection of all open subsets of $\mathbb{R}$ in the usual sense. Then $\mathcal{U}$ is a topology on $\mathbb{R}$.
(2) Let Y be any set and $\mathcal{U}$ the collection containing the empty set and all sets W such that $Y \setminus W$ is finite. Then $\mathcal{U}$ is a topology on Y.
(3) Let Y be any set and $\mathcal{U}$ the collection that has just two elements: Y itself and the empty set $\emptyset$. Then $\mathcal{U}$ is a topology.
(4) Let $\mathcal{V}$ be a topology on Y and let $f : X \to Y$ be a surjective function. Then
$$\mathcal{U} = \{U = f^{-1}(V) : V \in \mathcal{V}\}$$
is a topology on X.
(5) Let X be a set that is equipped with a metric d. Let $\mathcal{V}$ consist of the collection of all balls $B(p, r)$ determined by the metric and define $\mathcal{U}$ to be the collection of all sets obtained from $\mathcal{V}$ by finite intersection and arbitrary union. Then $\mathcal{U}$ is a topology on X.

A space X with topology $\mathcal{U}$ is said to be $T1$ if, for any two points $x, y \in X$, there is an open set $U \in \mathcal{U}$ and an open set $V \in \mathcal{U}$ such that **(i)** $x \in U$ and $y \notin U$ and **(ii)** $y \in V$ and $x \notin V$.

A space X with topology $\mathcal{U}$ is said to be *Hausdorff* if, for any two points $x, y \in X$, there are sets $U, V \in \mathcal{U}$ such that $x \in U$, $y \in V$, and $U \cap V = \emptyset$. Such a space is also called a $T2$ space.

A space X with topology $\mathcal{U}$ is said to be *regular* if, for any $x \in X$ and closed set $E \subset X$ with $x \notin E$, there are open sets $U, V \subset X$ such that $x \in U$, $E \subset V$, and $U \cap V = \emptyset$. If, in addition, each singleton set of the form $\{x\}$ is closed, then the space is said to be $T3$.

A space X with topology $\mathcal{U}$ is said to be *normal* if, for any closed sets $E, F \subset X$ with $E \cap F = \emptyset$, there are open sets $U, V \subset X$ such that $E \subset U$, $F \subset V$, and $U \cap V = \emptyset$. If, in addition, each singleton set of the form $\{x\}$ is closed, then the space is said to be $T4$.

If $(X, \mathcal{U})$ and $(Y, \mathcal{V})$ are topological spaces, then a mapping $\Phi : X \to Y$ is called *continuous* if, whenever $W \subseteq Y$ is an open subset, then $\Phi^{-1}(W)$ is open in X. The mapping is called a *homeomorphism* if Φ is one-to-one and onto and both Φ and Φ^{-1} are continuous.

Algebraic Topology

Let $(X, \mathcal{U})$ be a topological space. For $j = 1, 2$, let
$$\gamma_j : [0, 1] \to X$$

9.3. Geometry/Topology

be a closed curve. That is to say, each γ_j is a continuous function and $\gamma_j(0) = \gamma_j(1)$. We say that γ_1, γ_2 are *homotopic* if there is a continuous function $H : [0,1] \times [0,1] \to X$ such that

- The map $t \mapsto H(s,t)$ is a closed curve, each fixed $s \in [0,1]$;
- $H(0,t) = \gamma_1(t), \ 0 \leq t \leq 1$;
- $H(1,t) = \gamma_2(t), \ 0 \leq t \leq 1$.

We think of H as a continuous deformation of γ_1 to γ_2. We call H a *homotopy* and we say that γ_1 is homotopic to γ_2.

If $x \in X$ is a fixed point and if γ_1, γ_2 each begin and terminate at x, then we define a *fixed-point homotopy* of γ_1 to γ_1 to be one for which $H(s,0) = H(s,1) = x$ for all $0 \leq s \leq 1$.

Fix a point $x \in X$. Let $\mathcal{C}_x$ be the collection of closed curves in X which begin and terminate at x. If $\gamma_1, \gamma_2 \in \mathcal{C}_x$ then we write $\gamma_1 \sim \gamma_2$ provided that γ_1 is fixed-point homotopic to γ_2. Clearly $\sim$ is an equivalence relation. The collection of equivalence classes may be made into a group as follows: If $[\gamma_1], [\gamma_2]$ are equivalence classes and γ_1, γ_2 are representatives, then we set

$$[\gamma_1] \cdot [\gamma_2] = [\mu],$$

where

$$\mu(t) = \begin{cases} \gamma_1(2t) & \text{if } 0 \leq t \leq 1/2 \\ \gamma_2(2t-1) & \text{if } 1/2 < t \leq 1. \end{cases}$$

The resulting group is called the *first homotopy group* or *fundamental group*, denoted $\pi_1(X)$.

We say that the space X is *simply connected* if the first homotopy group is trivial (i.e., $\pi_1(X) = \{e\}$). The first homotopy group is, in general, not abelian. It is a device for measuring the presence of "holes" in the topological space X.

If $(X, \mathcal{U})$ and $(Y, \mathcal{V})$ are topological spaces and if $F : X \to Y$ is a continuous mapping, then there is an induced map

$$F_* : \pi_1(X) \longrightarrow \pi_1(Y)$$

$$[\gamma] \longmapsto [F \circ \gamma].$$

In case F is a homeomorphism then F_* is a group isomorphism.

The abelianization of the first homotopy group is the first homology group. This is another device of algebraic topology for measuring the presence of "holes". The dual of the first homology group is the first cohomology group.

Calculus on Euclidean Space

Let $U \subseteq \mathbb{R}^M$ and $V \subseteq \mathbb{R}^N$ be open sets. A mapping $F : U \to V$ is said to be *differentiable* at a point $P \in U$ if there is a linear mapping $\mathcal{L} = \mathcal{L}_P$ from $\mathbb{R}^M$ to $\mathbb{R}^N$ such that, for x near to P,

$$F(x) = F(P) + \mathcal{L}(x - P) + e_P(x).$$

Here the linear map $\mathcal{L}$ is applied to the M-vector $x - P$. Also the expression $e_P(x)$ is supposed to satisfy $\lim_{x \to P}[e_P(x)/|x - P|] = 0$. The mapping is *continuously differentiable* if the assignment $P \mapsto \mathcal{L}_P$ is continuous.

In practice, we calculate with partial derivatives. Let us write

$$F(x_1, x_2, \ldots, x_M) = \big(f_1(x_1, x_2, \ldots, x_M), \ldots, f_N(x_1, x_2, \ldots, x_M)\big).$$

Then, as we know from calculus,

$$\frac{\partial f_j}{\partial x_k}(x)$$
$$= \lim_{h \to 0} \frac{f_j(x_1, \ldots, x_{k-1}, x_k + h, x_{k+1}, \ldots, x_M) - f_j(x_1, \ldots, x_{k-1}, x_k, x_{k+1}, \ldots, x_M)}{h}.$$

We say that F is continuously differentiable, or C^1, on U if each of the functions $\partial f_j / \partial x_k$ exists on the open set U and is continuous. [This definition is equivalent to the one given in the last paragraph in terms of linear mappings. In fact, the entries of the matrix representation for $\mathcal{L}$ are just the partial derivative functions $\partial f_j / \partial x_k$ evaluated at P.] Inductively, the mapping F is C^m if $\partial f_j / \partial_k$ is C^{m-1} for each j, k. The mapping is C^∞ if it is C^m for every m.

The mapping $F : U \to V$ is said to be a C^m *diffeomorphism* if F is C^m, F is one-to-one and onto, and F^{-1} is C^m. In this circumstance it of course must be that $M = N$.

Manifolds

Let M be a second-countable (the topology has a countable basis), Hausdorff topological space. We say that M is a k-dimensional (Euclidean) manifold if, for each $m \in M$, there is a neighborhood W_m of m and a homeomorphism $\varphi_m : W_m \to E_m \subseteq \mathbb{R}^k$, where E_m is an open subset of the Euclidean space $\mathbb{R}^k$. The open set W_m is called a *coordinate patch* or *coordinate chart* at m and the mapping φ_m is called a *coordinate map*. Observe that φ_m endows the set W_m with k-dimensional Euclidean coordinates.

In geometry it is useful to mandate additional structure on the manifold. Let M be a manifold as in the last paragraph. Let $k > 0$ be an integer. We

9.3. Geometry/Topology

say that M is a C^k manifold if the coordinate charts have the property that

$$\varphi_{m'} \circ \varphi_m^{-1} : \varphi_m(W_m \cap W_{m'}) \to \varphi_{m'}(W_m \cap W_{m'})$$

is a C^k diffeomorphism. This definition also applies when $k = \infty$.

Let M be a C^k manifold. A function $f : M \to \mathbb{R}$ is said to be C^k if, whenever (W_m, φ_m) is a coordinate chart on M, then $f \circ \varphi_m^{-1}$ is a C^k function on $\varphi_m(W_m)$.

Let M be a C^1 manifold. Fix a point $P \in M$. We consider curves $\phi : (-1, 1) \to M$ such that $\phi(0) = P$. We say that two such curves ϕ and $\widetilde{\phi}$ are "equivalent" if, for each C^1 function $f : M \to \mathbb{R}$, it holds that

$$\frac{d}{dt}[f \circ \phi](0) = \frac{d}{dt}[f \circ \widetilde{\phi}](0). \qquad (*)$$

This is an equivalence relation and the equivalence classes are called the *tangent vectors* to M at P.

The collection of tangent vectors at $P \in M$ forms a vector space (just because the differentiation process is additive). We denote this space by $T_P(M)$ and call it the *tangent space* to M at P. The union of all of the tangent spaces, for $P \in M$, is called the *tangent bundle* for M. It is denoted by $\mathcal{T}M$. It is possible to topologize $\mathcal{T}M$. The topology and geometry of the tangent bundle can teach us a great deal about M, but we cannot explore that topic here (instead see, for example, [KON]).

In case M is the Euclidean k-space $\mathbb{R}^k$, then it is interesting to take $\phi_j(t) = (0, 0, \ldots, 0, t, 0, \ldots, 0, 0)$, where the t is in the jth position. Then the tangent vector corresponding to ϕ_j can be thought of as $\partial/\partial x_j$ in a natural way (just use the chain rule on $(*)$).

In case M is a C^1 manifold, and $P \in M$, then one can use a coordinate patch at P to express tangent vectors at P in terms of the local coordinates. We refer the reader to [LOS] for all of the details of this calculation.

Let M and N be C^k manifolds. A map $\Phi : M \to N$ is said to be a C^k mapping if, whenever (U_m, φ_m) is a coordinate chart on M and (V_n, ψ_n) is a coordinate chart on N, then $\psi_n \circ \Phi \circ \varphi_m^{-1}$ is a C^k mapping from Euclidean space to Euclidean space.

Riemannian Metrics

A *Riemannian metric* on a C^1, k-dimensional manifold M is an assignment

$$M \ni m \longmapsto (g_{ij}(m))$$

to each point $m \in M$ of a $k \times k$ positive definite matrix $\mathcal{G}(x) = (g_{ij}(x))$. The interest of the matrix $\mathcal{G}(x)$ is that it gives us a way to measure the lengths of tangent vectors v to M at a point $x \in M$. Namely, let W_x be a coordinate patch at the point x. Then, as discussed in the last subsection, the coordinate map allows us to realize v in coordinates. Then we set

$$\|v\|^2_{\mathcal{G},x} \equiv {}^t v \mathcal{G}(x) v \,.$$

Here the left superscript denotes the transpose of the vector v.

If the manifold M happens to be k-dimensional Euclidean space $\mathbb{R}^k$, then the standard Euclidean metric is obtained by taking $g_{ij}(x) = \delta_{ij}$ (the Kronecker delta) for every x. That is, the matrix $\mathcal{G}$ is the identity matrix. Then the (square of the) length of a tangent vector $\mathbf{v} = (v_1, \ldots, v_k)$ at the point $x \in \mathbb{R}^k$ is just

$$\|\mathbf{v}\|^2_{\mathcal{G},x} = \sum_{j=1}^{k} v_j^2 \,.$$

The interest of Riemann's concept of geometry is that we allow the definition of the length of a tangent vector to vary from point to point.

In most calculations involving a Riemannian metric it is convenient to assume that our manifold is C^2 and that the coordinates g_{jk} vary in a C^2 manner with the base point (so that we can calculate curvature).

The Picard Existence and Uniqueness Theorem

Suppose that $F(x, y)$ is a continuous function that satisfies a uniform Lipschitz condition in its second variable:

$$|F(x, y+h) - F(x, y)| \leq C|h| \,.$$

Then the initial value problem

$$\begin{aligned} \frac{dy}{dx} &= F(x, y) \\ y(x_0) &= y_0 \end{aligned}$$

has a unique solution in a neighborhood of the point (x_0, y_0).

This important and deep result of Picard has many applications in elementary differential geometry. For example, suppose that M is a smooth manifold and

$$M \ni x \longmapsto \mathbf{v}(x) \in T_x(M)$$

is a smooth assignment of a tangent vector at x to each $x \in M$. Such a mapping is called a *vector field*. A vector field may also be thought of as a smooth section of the tangent bundle.

Then the differential equation

$$\frac{dy}{dt} = \mathbf{v}(y(t))$$

is guaranteed, locally anyway, to have a unique solution. This gives rise to *integral curves* for the vector field. Here an integral curve is a smooth function $y = y(t)$ that has the property that its tangent vector at each point is the vector that comes from the given vector field.

The Inverse Function Theorem

Let
$$G : U \times V \to W$$
be a C^1 function, with $U \subseteq \mathbb{R}^M$ and $V, W \subseteq \mathbb{R}^N$. Assume that $O \in W$. Consider the equation
$$G(x, y) = 0.$$
We would like to know when we can solve this equation for y in terms of x—at least locally. The answer is that it is sufficient for the matrix
$$\frac{\partial G}{\partial y}$$
to have rank N.

This result, known as the *Implicit Function Theorem*, has important consequences for geometric analysis. First, if $F(x)$ is a smooth function on an open domain $U \subseteq \mathbb{R}^N$, if $S_c \equiv \{x \in U : F(x) = c\}$, and if $\nabla F \neq 0$ at points of S_c, then S_c is a smooth submanifold of $\mathbb{R}^N$.

Another interesting consequence of the Implicit Function Theorem is the *tubular neighborhood theorem*:

> **Theorem:** Let $M \subseteq \mathbb{R}^N$ be a smooth (at least C^2) hypersurface—for instance the boundary of a domain in $\mathbb{R}^N$. Then there is a neighborhood U of M such that each point of U has a unique nearest point in M, measured in the Euclidean metric.

A consequence of the tubular neighborhood theorem is that, in case M is compact, there is an $\epsilon_0 > 0$ such that the normal vectors to M of length ϵ_0 are disjoint. Thus they form an actualization of the normal bundle to M. The set M is an instance of a *set of positive reach*—see [FED].

Covariant Differentiation and Geodesics

For convenience now we restrict attention to a manifold that can be realized as a hypersurface M (i.e., a surface of dimension $(N-1)$) in Euclidean

N-space $\mathbb{R}^N$. Now we may represent a vector field in M as

$$X = a_1 \frac{\partial}{\partial x_1} + a_2 \frac{\partial}{\partial x_2} + \cdots + a_N \frac{\partial}{\partial x_N},$$

where the a_j are smooth functions on the surface M. If now X and Y are both vector fields, then we may define the *covariant derivative of Y in the direction X* at the point $p \in M$ to be

$$\overline{D}_X Y \equiv (X_p y_1, \ldots, X_p y_N),$$

where $Y = (y_1, \ldots, y_N)$ is the coordinate representation of Y and application of X_p simply means differentiation in the direction X at p. Observe that $\overline{D}_X Y$ can be calculated as soon as one knows Y along a curve that "fits" X, in the sense that the tangent to the curve at p is X_p.

Now let σ be a smooth curve in space, and let $T(p)$ be the tangent to σ at each point p. Let Y be a vector field. We say that Y is *parallel along σ* if $\overline{D}_T Y \equiv 0$ along σ. The curve σ is said to be a *geodesic* if $\overline{D}_T T \equiv 0$, that is, if the tangent field T is parallel along σ.

By Picard's theorem in differential equations, a geodesic is uniquely determined by **(i)** a point from which it emanates and **(ii)** its tangent at that point.

Although we have only discussed geodesics on hypersurfaces in Euclidean space, we may take it that they exist (locally) on any abstractly presented Riemannian manifold.

The Exponential Map

Let M be a C^2 manifold and fix a point $P \in M$. For each vector $\mathbf{v}$ in $T_P(M)$, Picard's theorem guarantees that there is a unique geodesic $\varphi_{\mathbf{v}}$ such that $\varphi_{\mathbf{v}}(0) = P$ and $\varphi'_{\mathbf{v}}(0) = \mathbf{v}$. The *exponential map* is defined to be the mapping

$$\mathbf{v} \longmapsto \varphi_{\mathbf{v}}(1).$$

The implicit function theorem can be used to show that this mapping, denoted $\exp_P$, is a local diffeomorphism of a neighborhood of 0 in $T_P(M)$ with a neighborhood of P in M. By transplanting coordinates from $T_P(M)$ to M, we can define local *geodesic normal coordinates* at P.

9.3. Geometry/Topology

A typical list of topics for the Geometry/Topology Qualifier is this:

- open, closed sets
- homeomorphisms
- homology
- differential calculus in $\mathbb{R}^N$
- smooth maps and their differentials
- rank theorem
- exterior derivative
- exact and closed forms
- differentiable manifolds
- forms and vector fields on a manifold
- orientation
- Stokes's theorem
- surfaces in $\mathbb{R}^3$
- Gaussian curvature
- the Levi-Civita connection
- geodesics and curvature
- geodesic convexity
- symmetric spaces
- the Lie derivative
- foliations
- universal covering space
- covering transformations
- van Kampen theorem
- fundamental group of a surface
- Riemann surfaces
- Lie groups and algebras
- 1-parameter subgroups
- closed subgroup theorem
- matrix groups ($SO(3)$, $SL(2,\mathbb{R})$, etc.)
- symmetric spaces
- homotopy
- vectors and k-forms
- inverse and implicit function theorems
- wedge product
- pull-backs of vectors and forms
- line integrals of 1-forms
- the tangent bundle
- manifolds with boundary
- partitions of unity
- Poincaré's lemma
- the Gauss map
- intrinsic geometry of surfaces
- Gauss-Bonnet theorem
- completeness and the Hopf-Rinow theorem
- spaces of constant curvature
- the flow of a vector field
- Frobenius theorem
- covering spaces
- path and homotopy lifting
- fundamental group
- classification of compact, oriented surfaces
- cohomology on surfaces
- de Rham cohomology and theorem
- correspondence of Lie algebra and group
- exponential map
- adjoint representation
- homogeneous spaces
- elementary representation theory
- separation axioms
- cohomology

9.4. Algebra

The subject of algebra is about *structures*. In graduate abstract algebra you are concerned with these structures:

- groups
- rings
- fields
- vector spaces
- modules

Serge Lang, in the venerable [LAN], posits the "monoid" (see below) as the basic unit on which all other structures are built. Many other references would begin with the group.

Groups

Of course a group is a set G equipped with a binary operation

$$G \times G \to G$$

which is usually denoted by juxtaposition or by $\cdot$. We refer to the binary operation as *multiplication* and we require that multiplication be associative. Further, we posit that there is an *identity* (or *unit*) element e such that $eg = ge = g$ for all elements $g \in G$. Finally we assume that, if $g \in G$, then there is an element $g^{-1} \in G$ such that $gg^{-1} = g^{-1}g = e$. We call g^{-1} the *inverse* of g.

Although many algebra books are rather short on examples, it is important for you (the student) to have many examples of groups (and also of the other basic structures in algebra) at your fingertips. Interesting examples of groups are these:

- Let S be a set. Then the collection of bijective mappings of S to S, equipped with the binary operation of composition, forms a group.
- Let T be a finite set. Then the collection of permutations of T forms a group under composition (this is a special case of the first example). The collection of permutations of *even order* (i.e., consisting of the composition of an even number of transpositions) forms a particularly important subgroup called the *alternating group*.
- The collection of $k \times k$ invertible matrices with real entries (usually denoted $GL(k, \mathbb{R})$ and called the *general linear group*) forms a group under matrix multiplication.

9.4. Algebra

- The set $\mathbb{Z}$ of integers, under the binary operation of addition, forms a group.
- The set of positive rational numbers, under the binary operation of multiplication, forms a group.
- A *cyclic group* G is one with the property that there is an $a \in G$ such that each element $g \in G$ has the form

$$g = a^n \equiv \underbrace{a \cdot a \cdots a}_{n \text{ times}}$$

for some nonnegative integer n.

A group G is called *abelian* if the group operation is commutative: $ab = ba$ for all $a, b \in G$. When the group is abelian, we often write the group operation as addition and the group identity as 0.

A *subgroup* H of the group G is a subset of G that is also a group, with the same identity element and the same binary group operation.

A subset $S \subseteq G$ *generates* G if each element of G can be obtained from elements of S by way of the group operations.

If G and $\widetilde{G}$ are groups, then a *group homomorphism* from G to $\widetilde{G}$ is a mapping $\varphi : G \to \widetilde{G}$ such that $\varphi(gg') = \varphi(g)\varphi(g')$. The mapping φ is called an *isomorphism* if it is one-to-one and onto, equivalently if there is a homomorphism $\psi : \widetilde{G} \to G$ so that $\varphi \circ \psi$ and $\psi \circ \varphi$ are the identity maps.

Now let G be a group and H a subgroup. A *coset* of H in G is a set

$$aH \equiv \{ah : h \in H\}$$

for some fixed $a \in G$. It is natural to ask when the collection of all cosets will form a group. The answer is that this happens precisely when H is a *normal subgroup*, which means that $x^{-1}Hx \subseteq H$ for all $x \in G$. The group operation is then $(aH) \cdot (bH) = (ab)H$. The group is called the *quotient group* and is denoted by G/H.

If G is a group with finitely many elements then we call the number of elements the *order* of G. If H is a normal subgroup then the cosets of H partition G into sets of equal size. So the order of H divides the order of G. The number of cosets of H in G is called the *index* of H in G and is denoted $(G : H)$.

If a is an element of G then the powers of a form a subgroup. We call the order of that subgroup the *order of a*. The subgroup is a cyclic group.

If $\mathbb{Z}$ is the group of additive integers, then let $p\mathbb{Z}$ denote the subgroup consisting of the multiples of the prime number p. Then $p\mathbb{Z}$ is a normal subgroup. We denote the quotient $\mathbb{Z}/p\mathbb{Z}$ by $\mathbb{Z}_p$ and call it the *cyclic p-group*. It is a fundamental result that any finite abelian group is isomorphic to a product of cyclic p-groups.

If $G, \widetilde{G}$ are groups and $\varphi : G \to \widetilde{G}$ is a homomorphism then the *kernel* of φ is given by
$$\ker(\varphi) = \{g \in G : \varphi(g) = e\}.$$
Of course the kernel is a normal subgroup of G. It follows that if $G = GL(k, \mathbb{R})$ (the general linear group) and the map φ is the determinant, then the kernel of φ is a normal subgroup. This kernel is $SL(k, \mathbb{R})$, the *special linear group* of order k.

If $\varphi : G \to H$ is a group homomorphism that is surjective then the *first fundamental isomorphism* of group theory says that
$$G/\ker(\varphi) \cong H,$$
where $\cong$ denotes group isomorphism.

A *tower* of groups is a sequence of subgroups
$$G = G_0 \supset G_1 \supset G_2 \supset \cdots \supset G_m.$$
We call the tower *normal* if each G_j is a normal subgroup of G_{j-1}. We say that the tower is *abelian* if it is normal and each G_{j-1}/G_j is an abelian group. We say that the top group G is *solvable* if it has an abelian tower whose terminal group is the trivial subgroup $G_m = \{e\}$. A basic fact is that if G is a group and H a normal subgroup then G is solvable if and only if H is solvable and G/H is solvable.

Sylow's fundamental theorem says that if the prime integer p divides the order of the finite group G then G has a subgroup of order p. More generally, if p^k is the highest power of p dividing the order of G then G has a subgroup of order p^k.

If $\{A_j\}_{j \in J}$ are abelian groups then we define their *direct sum*,
$$A = \bigoplus_{j \in J} A_j$$
to be the subset of the direct product $\prod_j A_j$ consisting of all tuples $(x_j)_{j \in J}$ with $x_j \in A_j$ and such that $x_j = 0$ for all but finitely many of the indices $j \in J$. Then A is a subgroup of the product. For each $j \in J$, define the map
$$\lambda_j : A_j \to A$$
by setting $\lambda_j(x)$ to equal x in the jth component and 0 otherwise. Then λ_j is an injective homomorphism.

9.4. Algebra

Proposition: Let $\{f_j : A_j \to B\}$ be a family of homomorphisms of the A_j into an abelian group B. Then there is a unique homomorphism
$$f : A \to B$$
such that $f \circ \lambda_j = f_j$ for all j. This is the *universal property* of the direct sum.

Let G be an abelian group. Let $\{e_j\}_{j \in J}$ be elements of G. We say that this family is a *basis* for G if each $g \in G$ may be uniquely written as
$$g = \sum \alpha_j e_j,$$
for $\alpha_j \in \mathbb{Z}$ and all but finitely many of the α_j equal to zero. An abelian group is *free* precisely when it has a basis. It follows then that $G \cong \bigoplus_j \mathbb{Z}_j$ where each $\mathbb{Z}_j$ is a copy of the integers $\mathbb{Z}$.

Now let S be any set. Let $\mathbb{Z}(S)$ be the set of all maps $\varphi : S \to \mathbb{Z}$ such that $\varphi(x) = 0$ for all but finitely many x. Then $\mathbb{Z}(S)$ is an abelian group under addition. We say that $\mathbb{Z}(S)$ is the *free group* generated by S and the set S composes its *free generators*.

Now let M be a set with a binary operation that is associative and has a unit (identity) element (such a structure is sometimes called a *monoid*). We may consider homomorphisms on M (i.e., maps that respect the binary operation and preserve the identity). Then there is a commutative group $K(M)$ and a homomorphism
$$\gamma : M \to K(M)$$
such that: If $f : M \to A$ is any homomorphism into an abelian group A, then there is a unique homomorphism $f_* : K(M) \to A$ such that
$$f = f_* \circ \gamma.$$
The group $K(M)$ is called the *Grothendieck group* of M.

Let A be an abelian group and assume that $mx = 0$ for each $x \in A$ and some fixed positive integer m. Let $\mathbb{Z}_m$ be the cyclic group of order m. Let $\text{Hom}(A, \mathbb{Z}_m)$ be the group of homomorphisms of A into $\mathbb{Z}_m$. Call this group the *dual* of A.

Rings

A *ring* is a set A together with two binary operations called addition and multiplication and denoted $+$ and $\cdot$. These are required to satisfy

R1. A is a commutative group under addition.

R2. Multiplication is associative and has a unit element.

R3. There are the distributive laws
$$(x+y)z = xz + yz \quad \text{and} \quad z(x+y) = zx + zy.$$

In a ring A, the unit element for addition is denoted 0 and the unit element for multiplication is denoted 1.

(1) The integers, with the usual laws of addition and multiplication, form a ring.

(2) The set of all $k \times k$ matrices with integer entries, with the usual laws of addition and multiplication, forms a ring.

(3) The set of all polynomials of a single variable x, with real coefficients and with the usual laws of addition and multiplication, forms a ring.

A subset B of the ring A is called a *subring* if it contains 0 and 1 and is itself a ring under the inherited ring operations.

If A is a ring, then let $U \subseteq A$ consist of those elements that have a right and left multiplicative inverse. Then U forms a group under multiplication that is called the *group of units*. We denote this group by A^*. In case $1 \neq 0$ and $A \setminus \{0\} = A^*$, then we say that A is a *division ring*.

A ring is said to be *commutative* in case the multiplication operation is commutative. A commutative division ring is called a *field*.

Now let G be a group and let K be a field. Denote by $K[G]$ the collection of all formal linear combinations $\alpha = \sum a_x x$, where $x \in G$ and $a_x \in K$; we assume that all but a finite number of the a_x are equal to zero. If $\beta = \sum b_x x$ is another such term, then we define the product

$$\alpha\beta = \sum_{x \in G} \sum_{y \in G} a_x b_y xy = \sum_{z \in G} \left(\sum_{xy=z} a_x b_y \right) z.$$

Under this binary operation, the set $K[G]$ becomes a ring, called the *group ring*. The second sum on the right is called a *convolution product*.

Let $\mathbf{a} \subseteq A$. We call the subset $\mathbf{a}$ of A a *left ideal* in the ring A if it is an additive subgroup and $A\mathbf{a} \subseteq \mathbf{a}$. For a *right ideal*, we instead require that $\mathbf{a}A \subseteq \mathbf{a}$. The ideal is *two-sided* if it is both a left and a right ideal. A two-sided ideal is called just an *ideal*. If $\mathbf{a} \subseteq A$ is an ideal and $\mathbf{a} \neq A$ then we say that $\mathbf{a}$ is a *proper ideal*. If A is a ring and $a \in A$ then the set Aa is a left ideal that we call a *principal ideal* (because it is generated by the single element a in an obvious sense). A commutative ring such that every ideal is principal is called a *principal ring* or a *principal ideal ring*. For example, the ring $\mathbb{Z}$ of integers is a principal ideal ring. [A similar definition of principal ideal may be given for *right multiplication*.]

9.4. Algebra

By a *ring homomorphism* we mean a mapping $\varphi : A \to B$ of rings such that
$$\varphi(a + \tilde{a}) = \varphi(a) + \varphi(\tilde{a}), \qquad \varphi(a\tilde{a}) = \varphi(a)\varphi(\tilde{a}).$$
It follows immediately that $\varphi(0) = 0$ and $\varphi(1) = 1$. The kernel of φ is of course the set of elements that are mapped to $0 \in B$. It is immediate that the kernel of a ring homomorphism is an ideal.

If $\mathbf{a}$ is an ideal in the ring A, then the *quotient ring* of A by $\mathbf{a}$ is simply the quotient calculated when A and $\mathbf{a}$ are viewed as additive groups. We denote this quotient by $A/\mathbf{a}$ and we endow it with binary operations by
$$(x + \mathbf{a}) + (y + \mathbf{a}) = (x + y) + \mathbf{a} \quad \text{and} \quad (x + \mathbf{a}) \cdot (y + \mathbf{a}) = xy + \mathbf{a}.$$
Then the *canonical map*
$$\begin{aligned} A &\longrightarrow A/\mathbf{a} \\ x &\longmapsto x + \mathbf{a} \end{aligned}$$
is a ring homomorphism.

A ring A is called an *integral domain*, or *entire*, if $1 \neq 0$, A is commutative, and there are no zero divisors in A (i.e., if $xy = 0$ then either $x = 0$ or $y = 0$).

Let A be a commutative ring. An ideal $\mathbf{p} \subseteq A$ is called *prime* if $\mathbf{p} \neq A$ and $A/\mathbf{p}$ is entire. The ideal $\mathbf{m} \subseteq A$ is called *maximal* if $\mathbf{m} \neq A$ and there is no larger proper ideal in A. Every maximal ideal is prime. Every proper ideal is contained in some maximal ideal. If $\mathbf{m}$ is a maximal ideal in A, then $A/\mathbf{m}$ is a field.

If $\mathbf{a} \subseteq A$ is an ideal and if $x, y \in A$ then we write $x \equiv y (\mathrm{mod}\, \mathbf{a})$ provided that x and y have the same image under the canonical map $A \to A/\mathbf{a}$.

> **Theorem [Chinese Remainder Theorem]:** Let A be a commutative ring. Let $\mathbf{a}_1, \ldots, \mathbf{a}_n$ be ideals of A such that $\mathbf{a}_j + \mathbf{a}_k = A$ for all $j \neq k$. If $x_1, \ldots, x_n \in A$, then there is an element $x \in A$ such that $x \equiv x_j (\mathrm{mod}\, \mathbf{a}_j)$ for all $j = 1, \ldots, n$.

Now we consider the abstract notion of a polynomial over a commutative ring A. Consider the free group generated by a single element X. This is an infinite cyclic group. Let S be the subset of this group consisting of powers X^r with $r \geq 0$. Then S is a *monoid* in the sense that it has an associative binary operation and a unit. The set of *polynomials with coefficients in A* is defined to be the set of functions $S \to A$ which are equal to 0 except for finitely many elements of S. Thus a polynomial can be written uniquely (with the group law now denoted by multiplication) as a finite sum

$$p(X) = a_0 X^0 + a_1 X^1 + \cdots + a_n X^n .$$

Here the a_j are elements of A and are called *coefficients* of the polynomial. The product is defined by the convolution rule, as discussed earlier. If A is a subring of a commutative ring B, then we can define the induced *polynomial function p* on B by

$$p(b) = a_0 + a_1 b + \cdots + a_n b^n$$

for $b \in B$.

We have only just begun a formal development of the graduate abstract algebra that is necessary for the qualifying exam. A typical, and fairly complete, list of all of the topics for the Algebra Qualifier is this:

- group theory
- solvability
- free abelian groups
- polynomial rings
- localization
- modules
- direct products of modules
- free modules
- dual spaces
- modules over principal rings
- polynomials over a factorial ring
- Hilbert's theorem
- symmetric polynomials
- field extensions
- splitting fields
- separable extensions
- inseparable extensions
- roots of unity
- norm and trace
- solvable and radical extensions
- integral ring extensions
- Galois cohomology
- alg. indep. of homomorphisms

- Sylow theorems
- direct sums
- free groups
- group rings
- principal and factorial rings
- homomorphisms of modules
- sums of modules
- vector spaces
- dual modules
- polynomials in one variable
- criteria for irreducibility
- partial fractions
- formal power series
- algebraic closure
- normal extensions
- finite fields
- Galois extensions
- linear independence of characters
- cyclic extensions
- abelian Kummer theory
- integral Galois extensions
- non-abelian Kummer extensions
- normal basis theorem

- infinite Galois extensions
- extensions of homomorphisms
- Hilbert's Nullstellensatz
- associated primes
- Nakayama's lemma
- ordered fields
- real zeros and homomorphisms
- dependence and independence
- finite extensions
- complex fields
- characteristic polynomials
- Jordan normal form
- flat modules
- tensor algebra of a module
- exterior algebra
- universal derivations
- homological algebra
- Grothendieck group
- homotopies of morphisms
- delta functors
- homology sequence
- the modular connection
- Noether normalization theorem
- Noetherian rings and modules
- primary decomposition
- discrete valuations
- real fields
- absolute values
- completions
- valuations
- polynomials in complete fields
- eigenvalues
- tensor products
- functorial isomorphisms
- symmetric products
- alternating products
- the de Rham complex
- Euler characteristic
- injective modules
- derived functors
- spectral sequences

9.5. How Do All of These Subjects Fit Together?

If you take a course in Banach algebras, then you will see how ring theory can give one an entirely new perspective on analysis and Banach space theory. If you take a course in von Neumann algebras, then you will see how the theory of groups and algebras can enrich the study of operator theory and mathematical physics. If you take a course in Lie groups, then you will see a beautiful interaction of analysis, differential geometry, group theory, and differential equations. If you take a course in analytic number theory, then you will see analysis, differential equations, and algebra interacting in surprising and productive ways. If you take a course in the hot new area of noncommutative geometry, then you will see how operator theory can be used to give an entirely new way to think about geometry.

One of the deepest and most exciting areas of modern mathematical research today is in the application of techniques of nonlinear partial differential equations to algebraic geometry. One of the most active areas of the modern theory of complex analysis is in the study of automorphism groups of domains and complex manifolds. The study of minimal surfaces and the Plateau problem uses techniques from geometry, differential equations, measure theory, homology, and hard analysis. Modern harmonic analysis uses ideas from geometry (curvature, geodesics, contact geometry, symplectic geometry, Euclidean geometry), differential equations, and measure theory. Modern mathematical physics uses deep ideas from sheaf theory, covering spaces, differential geometry, Lie groups, von Neumann algebras, and differential equations.

There is no way to learn of these complex, rich, and symbiotic relationships among the different parts of mathematics except by immersing yourself in them. And you cannot learn all of it. What you *can* hope to do is to become part of a group or a school, go to the seminars, read some of the papers, and begin to internalize the material gradually. It is a long, inductive process. After a few years of study, you will begin to speak the language. After a few more years of study, you will be able to formulate ideas and questions. After some more protracted study, you will begin to make your own contributions.

Part of maturing mathematically is to develop a point of view about the subject. This "angle" on things is your device for interpreting and absorbing new ideas. For example, Michael Atiyah interprets everything in terms of geometry. I know other mathematicians who interpret everything in terms of singular integrals, or in terms of categories. You know that you have really made mathematics a part of your makeup when you have strong opinions about the way it all fits together.

This is a profound, deep, and immensely rewarding process. It is the road to becoming a modern research mathematician. Seventy-five years ago, very few mathematical research papers were written collaboratively. Today, the majority of mathematics papers have more than one author. This is a beautiful manifestation of the fact that mathematics is a free and open subject in which we all share our ideas with joy and vigor. It is also a product of the fact that so much of the work is cross-disciplinary. We *must* collaborate in order to make any progress.

I should stress that the level of mastery that you must achieve in order to pass the quals does not involve any of the synthesis that I have just described. If you can internalize the topics as I have described them in Sections 9.1–9.4, then you will be in good shape for these exams. The sort of symbiosis and interaction that I am indicating in the present section is a long and protracted process that just begins while you are writing your thesis and that will unfold as your career progresses. It is something for you to look forward to.

No matter what graduate program you enter, strive to make the most of it. Take courses in many different fields—not just your narrow field of interest. Talk to students who are studying with other advisors, in other fields. Tell them about your thesis problem and listen to them discussing their problems. I am an analyst by training, but I have moved into the study of geometry and Lie groups and partial differential equations just because I like to talk to people about mathematics. I have learned from logicians and algebraists and geometers and PDE people and most anyone who would stand still long enough to talk to me. I have had dozens of collaborators and have learned a great deal from each of them. I would encourage you to do the same. It is a rich life, but it is a life that you make for yourself. It's not going to happen unless you make it happen. You immerse yourself in the mathematics and go with the flow. Let your interests and your passions lead you where they will. Learn what you can and make whatever contributions you can. That is the mathematical life.

Glossary

AAUP See *American Association of University Professors*.

A.B.D. Abbreviation for "all but dissertation" (also called "A.B.T.", or "all but thesis"). The phrase describes a student who has completed all parts of the Ph.D. program *except* for the dissertation. At many schools, "A.B.D." is an official status and you fill out some paperwork to ratify the fact that you have done everything but the dissertation. There are a great many students who leave graduate school at the A.B.D. stage and never complete the degree.

A.B.T. See *A.B.D.*.

academic integrity The rules of conduct by which we live academic life. These include not to cheat on exams, not to plagiarize, and to respect the work of others.

Academic Senate A governing body of the university, usually peopled by elected members of the faculty. Also called the *Faculty Senate*.

ACM See *Association for Computing Machinery*.

Acrobat A reader (created by Adobe Systems) for the `pdf` computer graphics and page design language. Particularly well-suited for use on the internet. See the URL
`http://www.adobe.com/products/acrobat/readstep2.html`.

actuary A mathematical scientist who calculates annuities, amortization plans, and other insurance data.

adjunct faculty Teaching faculty, usually those who are hired to teach specific, individual courses. Such people are paid by the course and usually have no benefits.

Adobe *Illustrator* Software, created by Adobe Systems for creating graphics on a computer. See the URL
`http://www.adobe.com/products/illustrator/main.html`.

American Association of University Professors (AAUP) A national professional organization for academic faculty. The AAUP protects the rights of faculty and often gets involved in legal cases (i.e., protests of tenure decisions, unlawful termination, etc.). See the URL `http://www.aaup.org`.

American Mathematical *Monthly* A primary mathematics journal of the Mathematical Association of America. See the URL
`http://www.maa.org/pubs/monthly.html`.

American Mathematical Society (AMS) One of several professional mathematical associations in the United States. Located in Providence, Rhode Island, the AMS was formed in 1888. The original purpose of the AMS was to concentrate on research concerns, but it has in recent years expanded its purview. Consult the URL `http://www.ams.org`.

AMS See *American Mathematical Society*.

AMS Cover Sheet A standard information sheet, available on the internet and also in issues of the *Notices of the American Mathematical Society*, to be included with job application materials.

GLOSSARY

AMS Employment Center The job interview activities sponsored by the AMS/MAA/SIAM at the January annual meeting.

$\mathcal{AMS}$-LaTeX A version of TeX, the computer typesetting language. See the URL `http://www.ams.org/tex/amslatex.html`.

$\mathcal{AMS}$-TeX A version of TeX, the computer typesetting language. See the URL `http://www.ams.org/tex/amstex.html`.

American Statistical Association (ASA) One of several professional mathematical associations in the United States. The purpose of the ASA is to support and promote statistical activities and scholarship. Consult the URL `http://www.amstat.org`.

application deadline Most graduate programs have a definite deadline by which all applications must be submitted.

application essay Most graduate school applications ask you to write a brief essay about your career goals and why you wish to attend graduate school.

application fee There is usually a fee, of $40 or $50, to apply to a graduate program.

ASA See *American Statistical Association*.

ASL See *Association for Symbolic Logic*.

Assistant Professor This is an academic member of the department, on the tenure track. The Assistant Professor does not have tenure, but is a candidate for tenure.

Associate Professor This is an academic member of the department, on the tenure track. The Associate Professor usually has tenure.

Association for Computing Machinery (ACM) A national organization that is "a major force in advancing the skills of information technology professionals and students." The Association engages in publishing and organizes conferences. See the URL http://www.acm.org.

Association for Symbolic Logic (ASL) A national mathematical organization that concerns itself with fostering and promoting logic and issues that are of concern to logicians. The Association publishes journals and books and organizes conferences. See the URL http://www.aslonline.org.

Association for Women in Mathematics (AWM) A national mathematical organization that concerns itself with promoting the role of women in mathematics. See the URL http://www.awm-math.org/.

attrition rate The percentage of people who drop out of a program before completion.

AWM See *Association for Women in Mathematics*.

Bulletin of the American Mathematical Society A primary journal of the American Mathematical Society. See the URL http://www.ams.org/bull/.

Carnegie Classification of Institutions of Higher Education The leading typology of American colleges and universities. Provides useful rankings and comparisons. Originally published in 1973, there have been updated editions in 1976, 1987, 1994, and 2000. The centennial edition will be published in 2005. Consult the URL
http://www.carnegiefoundation.org/Classification.

CBMS See *Conference Board of the Mathematical Sciences*.

CBMS Conference A week-long conference, usually held at a nonresearch college or university, which features a prominent mathematician giving ten lectures (two per day). There are also about ten lectures by other mathematicians. These events, which are held every year, are sponsored by the CBMS.

Chairperson The titular director of the Department. Usually the Chairperson is appointed by the Dean, with the approval of the Department.

Chair Professor See *Endowed Chair Professor*.

coffee room A room set aside in the math department for getting coffee, for socializing, and for talking about mathematics. Often the mailboxes are located in the coffee room. The pre-colloquium tea is usually held in the coffee room.

collaborator Another mathematician or scientist who will work with you on a research or authoring project.

colloquium A formal lecture, usually given by a member of another department and often by a professor from another university, that is given for the benefit of the entire math department. The lecture is usually preceded by a ceremonial tea and there is often a celebratory dinner and even a party afterward.

comprehensive university These institutions started out as the "normal schools", that is, schools that were dedicated to teacher training. Seventy-five years ago there were hundreds of these throughout the country. Today most of these institutions have changed their names, and in some cases, their missions.

Concerns of Young Mathematicians (CYM) An internet periodical devoted to issues that are of interest to those beginning in the mathematics profession. See the URL http://www.youngmath.net/concerns.

Conference Board of the Mathematical Sciences (CBMS) A national board, a subsidiary of the National Academy of Sciences, that oversees the welfare of the mathematical enterprise in this country. Of particular note are the CBMS conferences. See the web sites http://www7.nationalacademies.org/bms/ and http://www.cbmsweb.org.

Corel *DRAW* Software for creating graphics on a computer. See the URL http://www.corel.com.

course requirement Most graduate programs require that the student take a certain number of credit hours of courses.

Curriculum Vitae (CV) The analogue of what business people call a résumé. This document provides your personal and professional information. It is part of any job application that you may submit.

CV See *Curriculum Vitae.*

CYM See *Concerns of Young Mathematicians.*

DARPA See *Defense Advanced Research Projects Agency.*

Defense Advanced Research Projects Agency (DARPA) A branch of the CIA (Central Intelligence Agency) that is engaged in high-level technical research—often related to espionage. A generous source of funding for some mathematicians. See the URL `http://www.darpa.mil`.

Department Administrator See *Office Manager.*

departmental service Service by faculty on departmental committees, engaging in teaching evaluation, working on curriculum development, and many other activities as well. Departmental service (and university service as well) figures into all tenure and promotion decisions.

Department of Energy (DOE) An agency of the federal government that is concerned with energy issues and research into parts of science that impact on energy. In recent years, DOE has been a significant source of funding for mathematical research. See the URL `http://www.eia.doe.gov`.

dissertation Also called the *thesis.* The *magnum opus* of a Ph.D. program, this document (often 75 pages or more) is the student's disquisition on original research.

Doctor of Philosophy See Ph.D.

Doctor of Science See D.Sc.

DOE See *Department of Energy*.

D.Sc. An expository doctoral degree. The thesis involves no research.

EIMS See *Employment Information in the Mathematical Sciences*.

elite private university These are generally privately funded institutions with large endowments. They receive no funding from the state or federal government and are therefore quite independent in their policies and educational procedures.

Employment Information in the Mathematical Sciences (EIMS) A hard-copy publication of the mathematical societies in which current job openings are advertised. See the URL http://www.ams.org/eims/eims-search.html.

Endowed Chair Professor A Professor for whom the salary, travel funds, and other perks of the position comes from a special, endowed fund. It is a great honor to be an Endowed Chair Professor.

English as a Second Language (ESL) An academic program to help nonnative English speakers become proficient in the language.

ESL See *English as a Second Language*.

fellowship A grant to pay some or all expenses and support for either a graduate student or a faculty member.

Faculty Senate See *Academic Senate*.

Fields Medal Arguably the highest encomium that can be awarded a (young) mathematician. Established with money from John Charles Fields, the first Fields Medals were awarded in 1936 (to Lars Ahlfors and Jesse Douglas). Fields Medals are awarded every four years at the International Congress of Mathematicians. From two to four medals are awarded on each occasion. By custom (but not by mandate), Fields Medals are only awarded to mathematicians who are under the age of forty. See the URL http://www.mathunion.org/medals/.

final oral   See *thesis defense*.

financial statement   An attestation of financial resources that undergraduates applying for financial aid must complete. Such paperwork is usually not required of graduate school applicants.

"fired"   The mechanism by which a thesis advisor declines to continue directing a Ph.D. student's thesis.

foreign language requirement   Most Ph.D. programs in mathematics require that the student demonstrate reading knowledge of one or two foreign languages. French, German, and Russian are the most common languages that programs will consider. Usually the student is required to translate a page from some standard foreign math text.

four-year teaching college   An institution of higher education which has no Ph.D. program and often no Master's programs either. Such an institution usually places a greater emphasis on teaching than on research (in its promotion and tenure decisions, for example). The four-year teaching college sees its primary function to be undergraduate education.

generals   See *qualifying exams*.

Ghostscript   A useful utility for reading PostScript files. See the URL http://www.cs.wisc.edu/~ghost/.

Graduate (Admissions) Committee   The Math Department committee that oversees admission of graduate students and helps the Graduate Director run the graduate program.

Graduate Chairperson   See *Graduate Director*.

Graduate Director   The Graduate Director is part of the administrative structure of a university mathematics department (see Appendix I for more on administrative structure). S/he is in charge of seeing that the graduate program—including admissions and mentoring of the graduate students—runs smoothly. The Graduate Director works alongside the Chairperson and other members of the departmental administration to see that the Mathematics Department sticks to its agenda and accomplishes its goals. The Graduate Director is usually assisted by a Graduate Committee.

Graduate Faculty Those faculty who are qualified to direct a Ph.D. thesis.

Graduate Record Exam (GRE) A standardized screening test, administered by the Educational Testing Service of Princeton, New Jersey, to determine eligibility for graduate school. See the URL `http://www.gre.org/splash.html`.

graduate school An educational program that follows upon the usual four-year American undergraduate educational experience. Among other degrees, the graduate program will offer the Master's degree, the Ph.D., or both.

Graduate Vice Chairperson See *Graduate Director*.

Harvard *Graphics* Software for creating graphics on a computer. See the URL `http://www.harvardgraphics.com/index.asp`.

Head The titular director of the Department. Usually a Head is more autonomous than a Chairperson. The Head is appointed by the Dean.

Humboldt Foundation Fellowship A grant from the Alexander von Humboldt Foundation to enable a non-German scholar to undertake a long-term period of research in Germany. See the URL `http://www.ats.edu/faculty/external/spons/H0000448.html`.

ICM See *International Congress of Mathematicians*.

IDA See *Institute for Defense Analyses*.

IEEE See *Institute of Electrical and Electronics Engineers*.

IMS See *Institute of Mathematical Statistics.*

IMU See *International Mathematical Union.*

Institute for Defense Analyses (IDA) A federal institution, with locations in Alexandria, Princeton, and San Diego, for studying issues (most of them mathematical) pertaining to the nation's defense systems. See the URL http://www.ida.org.

Institute of Electrical and Electronics Engineers (IEEE) This organization is the "leading authority in technical areas ranging from computer engineering, biomedical technology and communications, electric power, aerospace, and consumer electronics." The Institute engages in publishing activities and organizing conferences. See the URL http://www.ieee.org.

Institute of Mathematical Statistics (IMS) One of several professional mathematical associations in the United States. The purpose of the IMS is to support and promote statistical activities and scholarship. Consult the URL http://www.imstat.org.

Instructor Full-time academic teaching staff, but not tenure-track. The Instructor is paid a salary and enjoys full benefits.

International Congress of Mathematicians (ICM) An international mathematical event, held every four years, to celebrate and assess recent progress in mathematics. The Fields Medals are awarded at the ICM. The last ICM was held in Beijing and the next will be in Madrid. See the URL http://www.icm2002.org.cn.

International Mathematical Union (IMU) This is the international aggregation of mathematical scholars—in effect the union of all of the national mathematical societies. The IMU considers broad issues that affect mathematicians worldwide. It organizes and holds the International Congress of Mathematicians every four years (the last was held in Beijing and the next will be in Madrid). It awards the Fields Medals. See the URL http://elib.zib.de/IMU/.

internet application Many graduate programs will allow you to apply via the internet. Some will waive their fees for students who do so.

journal A periodical publication in which mathematical research is published. There are also journals that are devoted to teaching and to exposition.

junior college A college with a two-year curriculum leading to an Associate of Arts (or A.A.) degree. Junior colleges do *not* grant Bachelor's degrees. Junior colleges often act as feeder schools to the big state universities.

L$\mathcal{A}_{\mathcal{M}}$S-TEX A version of the computer typesetting system TEX.

large state universities Ever since the early 1960's (and, in some cases, much earlier than that), every state has had a well-developed system of publicly supported universities. In many states these are very large institutions. For families of modest means, the state university is the default place to send their children for higher education.

LATEX A version of TEX, the computer typesetting language. See the URL http://www.latex-project.org/.

leave See *unpaid leave*.

letters of recommendation For admission to graduate school, you will need at least three letters (of recommendation) from people who can attest to your mathematical abilities, your maturity, your balance and judgment, and your capacity for hard work. Such letters are also required when you seek a job.

M.A. See *Master of Arts Degree*.

MAA See *Mathematical Association of America*.

Major Oral A lecture given by a graduate student in preparation for thesis work. See *Minor Oral*.

Maple A computer algebra system that has powerful graphing and symbol-manipulation capabilities. See the URL http://www.waterloomaple.com/.

Master of Arts Degree A first-level graduate degree. Earning of this degree may entail passing a course requirement, the taking of some qualifying exams, and possibly writing an expository thesis.

Master of Science Degree A first-level graduate degree. Earning of this degree may entail passing a course requirement, the taking of some qualifying exams, and possibly writing an expository thesis.

Mathematica A computer algebra system that has powerful graphing and symbol-manipulation capabilities. See the URL
http://www.wolfram.com.

Mathematical Association of America (MAA) One of several professional mathematical associations in the United States. Located in Washington, D.C., the MAA was formed in 1916. The purpose of the MAA is to concentrate on teaching concerns and the exposition of mathematics. Consult the URL http://www.maa.org.

Mathematical Reviews The hard-copy archiving tool of the American Mathematical Society, extant since 1940. This periodical records all papers published in all of the major journals, together with complete bibliographic information and a concise review. See the URL
http://www.ams.org/mr-database.

MathSciNet An internet tool, created by the AMS, that indexes mathematical papers published in hundreds of mathematics journals around the world. Each paper is carefully reviewed and the bibliographic reference provided. *MathSciNet* is the digital-age version of *Mathematical Reviews*. See the URL http://www.ams.org/mathscinet/.

MiKTeX A freeware version of TeX, available on the internet. See the URL http://www.miktex.org.

Minor Oral A lecture given by a graduate student in preparation for thesis work. See *Major Oral*.

Mitre Corporation A high-tech firm, founded by mathematician John Kemeny and others, that hires many mathematicians. Much of the work of Mitre is in defense applications of mathematics. See the URL http://www.mitre.org/.

M.S. See *Master of Science Degree*.

National Research Council (NRC) A federal organization that oversees research programs for the government. See the URL http://www.nas.edu.

National Research Council (NRC) Group Rankings The National Research Council's ranking of mathematics Ph.D. programs entails, among other things, a grouping of mathematics Ph.D. programs into "Group I", "Group II", ..., up to Group V. Departments are ranked according to several characteristics, the main one being the scholarly quality of the faculty. In the 1995 ranking, Group I comprises 48 departments with scores (in the report [GMF]) in the range 3.00–5.00. Group II comprises 56 departments with scores in the range 2.00–2.99. Group III comprises 72 departments with scores in the range 0–1.99. Group IV lists doctoral programs in statistics, biostatistics, and biometrics. Group V lists doctoral programs in applied mathematics and applied science.

National Science Foundation (NSF) The primary source of research funding for mathematicians—especially pure mathematicians. See the URL http://www.nsf.gov.

National Security Agency (NSA) A federal agency that works in mathematical areas pertaining to the security of the nation. A particular specialization is cryptography. In addition to being a likely place of employment for Ph.D. mathematicians, the NSA is also a source of research funding. See the URL http://www.nsa.gov.

normal progress The prescribed rate, in a graduate program, at which one should take the qualifying exams, pass courses, get a thesis advisor, and write a thesis.

normal school See *comprehensive university*.

Notices of the American Mathematical Society A primary journal of the American Mathematical Society. This is *not* a research journal. Rather, it is the official organ of the AMS. It contains a great deal of society news. See the URL http://www.ams.org/notices.

NRC See *National Research Council.*

NSA See *National Security Agency.*

NSF See *National Science Foundation.*

NSF CAREER Award Part of the Faculty Early Career Development Program of the National Science Foundation. This program is the NSF's most prestigious award for new faculty members. It supports the early career-development activities of those teacher-scholars who are most likely to become the academic leaders of the 21st century. See the URL http://pag.lcs.mit.edu/~mernst/advice/career-grant.html.

NSF Postdoctoral Fellowship A fellowship program of the National Science Foundation designed to permit participants to choose research environments that will have maximal impact on their future scientific development. The program provides 24 months of support for each awardee. See the URL http://www.math.uiowa.edu/~tomforde/NSFpostdoc.html.

Office Manager The senior staff person in the department, sometimes called the *Department Administrator.*

orientation A period at the beginning of the school year during which beginning graduate students are introduced to the program, to their duties, and to the activities of graduate students.

***pdf* format** A file format (Portable Document File format) due to Adobe. Files in this format generally have extension .pdf, as in myfile.pdf. Now a standard file format for electronic document printing and distribution. The Adobe Acrobat reader is designed to read *.pdf files. Files in the *.pdf format tend to be smaller than the corresponding PostScript (or *.ps) files—especially for text. See the URL
http://www.adobe.com/products/acrobat/readstep2.html.

Ph.D. The Doctor of Philosophy, highest academic degree granted by most universities. Earning of a Ph.D. involves (at least) the passing of qualifying exams, fulfillment of a course requirement, and the writing of a thesis containing original research.

Ph.D. Candidate A student who has passed the qualifying exams and has a thesis advisor and thesis problem.

Ph.D. thesis See *thesis*.

PostScript A page design language created by Adobe Systems. This is a high-level computer language that allows the formatting of both graphics and text. See the URL
http://www.adobe.com/products/postscript/main.html.

prelims See *qualifying exams*.

professional society The professional societies (AMS, AWM, MAA, SIAM, etc.) are collegial organizations of professional mathematicians. Their purpose is to support and promote research and teaching in mathematics.

Professor This is an academic member of the department, on the tenure track. The Professor has tenure and is a senior member of the department. In most academic departments, there is no rank higher than Professor.

Project NExT A consortium of young mathematicians across the country that gathers regularly to share concerns and to help bring beginners in the profession up to speed. Project NExT particularly stresses networking among its participants. See the URL
http://archives.math.utk.edu/projnext/.

publishing a paper The process by which one writes up a paper with original results, submits it to a journal, undergoes the refereeing process, and then has the work appear in a journal.

"Publish or perish." A byword of American higher education for the past century, this phrase encapsulates the sentiment that one must engage in academic research and publishing in order to obtain tenure and to flourish as a professor.

qualifying exams A set of examinations, usually a standard part of any Ph.D. program, to determine whether the student has the background and ability to go on and write a Ph.D. thesis.

quals See *qualifying exams*.

R.A. See *Research Assistant*.

RAND Corporation A think tank, located in Santa Monica, California, that consults for the Federal Government. See the URL http://www.rand.org.

research The scholarly pursuit of endeavoring to discover new mathematics, or new facts and theorems about existing areas of mathematics. Research is usually published in journals, or sometimes in books.

Research Assistant (R.A.) A form of graduate student financial support in which the student duties consist of assisting a professor in research activities.

Research Faculty See *Graduate Faculty*.

research grant A source of funding to support research. A research grant can pay for travel, for visits from mathematical collaborators, for the support of graduate students, and to buy equipment. Some research grants pay the principal investigator a summer salary. Common sources of funding for grants are the National Science Foundation, the National Security Agency, DARPA, the Navy, the Department of Energy, the Department of Education, and the Army. Private corporations or foundations sometimes also provide grants.

Research Instructor The Research Instructor is an instructor in an honorary position usually paid by a special endowment. Often the Research Instructor has a reduced teaching load and a special travel fund. Research

Instructorships exist primarily, but not exclusively, at the elite private universities.

Research Statement Often it is expected of a young mathematician applying for an academic position to supply a statement of research accomplishments and future plans. This document may be two to three pages in length.

research university An educational institution that grants undergraduate degrees (the B.A. and/or the B.S.) and also graduate degrees (the M.A., M.S., and Ph.D.). Such an institution will also have a faculty that has a vigorous research program, publishes widely, and visits universities all over the world (to collaborate with their faculties and to lecture about their work).

résumé The version of the *Curriculum Vitae* or *CV* that is used in the business world.

sabbatical Many institutions of higher education, especially research universities, offer to their faculty the opportunity to take a paid leave of absence every seven years. At some schools this privilege is guaranteed. At others one must compete for a sabbatical.

seminar This is a working group that usually meets once a week to learn some focused subject area of mathematics. Most of the time, the seminar speakers will be members of the home math department.

SIAM See *Society for Industrial and Applied Mathematics.*

SIAM News A primary journal of the Society for Industrial and Applied Mathematics.

SIAM Review A primary journal of the Society for Industrial and Applied Mathematics.

Sloan Foundation Fellowship Fellowships given to enhance the careers of the very best young faculty in the sciences. The Sloan Fellowship enables a young mathematician to take a one-year (or longer) leave of absence at another research institution. See the URL `http://www.sloan.org`.

Society for Industrial and Applied Mathematics (SIAM) Located in Philadelphia and founded in 1952, this professional mathematical association nurtures and promotes the concerns of mathematicians who deal with applications. See the URL http://www.siam.org.

Sputnik era A period of ten years in the United States, following the launch in 1957 of the Soviet satellite *Sputnik*, of rapid development in mathematics and science education.

staff These are the nonacademic members of the math department, including secretaries, computer managers, some counselors, and so forth.

Statement of Purpose See *application essay*.

Steele Prize There are three prizes named after Leroy P. Steele, who set up the fund that endows the awards. One prize is for lifetime achievement in mathematics, one prize is for a book or substantial survey in mathematics, and one prize is for a paper of lasting and fundamental influence. The American Mathematical Society administers the Steele Prizes. See the URL http://www.ams.org/prizes-awards.

state universities See *large state universities*.

TA See *Teaching Assistant*.

TA Training Most math departments provide some lessons or discussion to help new graduate students understand the duties of the TA (Teaching Assistant) and what teaching skills will be required.

Teaching Assistant (TA) A graduate student who assists in the teaching of university classes. This is usually done as part of the arrangement to justify the financial support of a graduate student.

teaching load The number of courses per semester, or per year, that you are required to teach. Mathematicians frequently describe their teaching load as "3-and-2"—meaning that they teach three courses in one semester and two courses in the other. Sometimes people will describe their teaching load in terms of the number of contact hours.

Teaching Statement A brief essay, usually required of junior people who are applying for academic employment, in which the candidate describes teaching experience, values, and goals.

tenure Tenure is a means of giving academic professors job security, protection of their time and discretion to study a broad range of subjects, and protection against societal attack. Tenure is a lifetime privilege, difficult to obtain and difficult to lose.

tenure document A document, with legal standing, that defines and formulates the university's policies toward faculty tenure.

tenure dossier The collection of materials, including teaching evaluations, outside review letters, information about service, a publication list, and other information that pertains to a tenure case.

tenure-track Describes those positions in the department that may lead to tenure.

Test of English as a Foreign Language (TOEFL) This exam is given to nonnative English speakers to determine their proficiency in the language. See the URL http://www.toefl.org.

TeX A computer typesetting system for creating mathematical documents. See the URL http://www.tug.org for the TeX Users Group.

TeX Users Group (TUG) The official organization for the development and promotion of the computer typesetting system TeX.

thesis Also called the *dissertation*. The *magnum opus* of a Ph.D. program, this document (often 75 pages or more) is the student's disquisition on original research.

thesis advisor The professor who directs the Ph.D. thesis. Often the advisor provides the problem on which the student works, and offers advice and encouragement along the way. The thesis advisor will be a tenure-track faculty member.

thesis committee The committee, chaired by the thesis advisor and composed of members of the math department plus select members of other departments, that adjudicates the validity of the Ph.D. thesis. The thesis committee presides over the *thesis defense*.

thesis defense The final ceremonial presentation by the Ph.D. candidate of the thesis results to the Ph.D. committee and a select audience.

thesis problem The question or subject area that you will study in order to develop the materials for your Ph.D. thesis.

TOEFL See *Test of English as a Foreign Language*.

TUG See the *TeX Users Group*.

undergraduate research A sort of research activity that some undergraduates perform under the close tutelage of a faculty mentor.

undergraduate school The standard, four-year college curriculum that follows directly after high school.

underrepresented group A societal group—women, African-Americans, Native Americans, or one of several others—that has a lower percentage of representation in the mathematics profession than the percentage of its representation in the American population.

university service Service by faculty on university-wide committees. This may entail teaching evaluation, curriculum development, evaluating tenure cases, administering discipline, or many other duties. University service (and departmental service as well) figure into all tenure and promotion decisions.

unpaid leave This is a leave of absence from your home university or institution in which you are paid entirely by the host institution (and your home institution provides no funds).

Vice Chairperson for Graduate Studies See *Graduate Director*.

vita   See *Curriculum Vitae*.

Daniel Wagner Associates   A private corporation, founded by mathematicians, that does consulting for the government and for industry. One specialty of Daniel Wagner is operations research. See the URL http://www.wagner.com/site/main.html.

Wolf Prize   The Wolf Prize is awarded each year in the fields of agriculture, chemistry, mathematics, medicine, physics, and the arts. Administered by the Wolf Foundation in Israel, and named after Ricardo Wolf, the prize has become one of the top encomia in mathematics. See the URL http://aquanet.co.il/wolf/wolf5.html.

xfig   UNIX software for creating graphics on a computer. See the URL http://www.xfig.org.

Zentralblatt für Mathematik   The European version of MathSciNet, published by Springer-Verlag. Their digital-age, online version is available for free on the web at http://www.emis.de/ZMATH. See also http://www.zblmath.fiz-karlsruhe.de/.

APPENDIX I: The Administrative Structure of a Mathematics Department and a University

The Board of Trustees (private school) or Board of Regents (public school) usually oversees the running of the university at a very high level. These boards are usually peopled by high-ranking members of the community—businesspeople, civic leaders, and the like. Important alumni can be elected to the Board of Trustees/Regents.

A University is usually run by a Chancellor or a President. The Chancellor or President is usually answerable to the Board of Trustees/Regents. Such a person is the figurehead for the institution. A large part of this person's duties is to interface with the public and with the state or city government (especially if the institution is a state or city university). The Chancellor, generally speaking, is *not* primarily concerned with the day-to-day operation of the university. His/her main concerns include the budget and fundraising.

The Chancellor is assisted by many Executive Vice Chancellors, Vice Chancellors, Associate Chancellors, and Assistant Chancellors. My university has 42 of these people, and that number is modest compared to the cognate number at some of the big schools. Some of the purviews of these Associates and Assistants are:

- Medical Affairs
- Legal Matters
- Research
- Finance
- Clinical Affairs
- Undergraduate Affairs
- Alumni and Development
- Information Systems
- Academic Affairs
- Operations
- Veterinary Affairs
- Freshman Transition
- Publication
- Library Technology

Well, you get the idea. The modern university is a complicated place. It takes a lot of people to keep it running smoothly.

The person who *is* in charge of the day-to-day operations is the Provost. The Provost runs the university. S/he is in charge of tenuring and hiring faculty, of student discipline problems, of faculty discipline problems, and of broad policy issues affecting the university's people.

The Provost is assisted by various Associate Provosts and Assistant Provosts, and also by a variety of Deans. There are usually (at the very least):

- A Dean of Admissions
- A Dean of Undergraduate Studies
- A Dean of Graduate Studies
- A Dean of Arts & Sciences
- A Dean of Engineering
- A Dean of the Business School
- A Dean for Minority Affairs
- A Dean of Scholarships and Financial Aid

- A Dean of Capital Projects
- A Dean of Sports

It should be stressed that these Deans are not all co-equal. The Dean of Arts & Sciences will be of preeminent importance (for example, this Dean plays a central role in the hiring and tenuring of faculty); the Dean of Sports perhaps less so.

Ever since the 1960's, the dawn of the era of the mega-universities, university administrations have mushroomed and proliferated. There are a terrifying number of Assistant Deans and Associate Deans and Vice Provosts. Figure 1 gives an outline of the (skeleton of the) administration of a typical university. Of course nothing is etched in stone and you will not find this literal structure at every university. For example, Princeton has no Provost; instead it has a Dean of the Faculty. My own university has no Provost because our Chancellor plays a dual role as both Chancellor and Provost (and our Dean of Arts & Sciences plays the dual role of Dean and Provost).

Under the Deans and the Assistant and Associate Deans are the Department Chairs. Each academic department has a Chairperson. Sometimes instead the Dean prefers to have a Head. The basic difference between a Chairperson and a Head is that a Headship is a long-term appointment and the Head is fairly autonomous. The Head can make many decisions without a faculty vote. A Chair, instead, is a peer leader among his faculty. The charming and energetic book [CON1] discusses this distinction in some detail.

Usually a given academic department, say a math department, has a vote to determine who its new Chairperson will be. This is done every three to five years, depending on the agreed-upon term of office. It should be clearly understood that, at the vast majority of American universities, *the department does not choose its own Chair*. It is the Dean who selects and appoints the Chairperson, but s/he does take the Department's mandate into consideration. It is always best if the Dean can appoint the Chairperson whom the Department prefers.

The Chairperson usually is a full Professor. This is because the Chairperson will oversee the promotion of Assistant Professors to Associate Professorships and also the promotion of Associate Professors to Professorships. It would be awkward for an Assistant or Associate Professor to adjudicate the promotions of his peers. Also the Chairperson must have some clout, both within the department and around the university, so it is best to have a senior person in the job.

Finally, a Math Department Chairperson will ask certain of his colleagues to assist him/her in running the department. Often the Chairperson

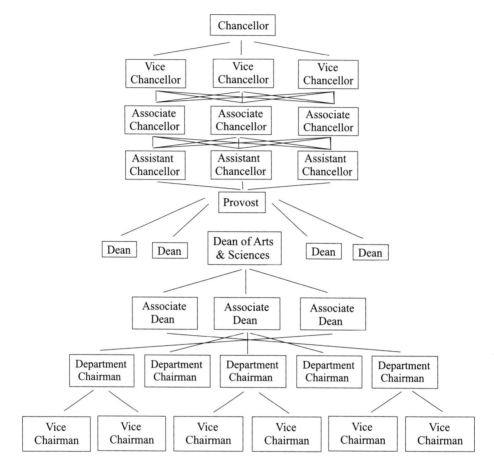

Figure 2. Administrative Structure of a University

will ask one person to be the Vice Chairperson for Undergraduate Studies and another to be the Vice Chairperson for Graduate Studies.[1] In many departments there is also an Administrative Vice Chairperson. These roles are simply a device for the Chairperson to delegate tasks and duties. Running the Ph.D. program and running the majors program are big jobs. It is useful to have good people to help out with these tasks.

Any university or college is replete with committees. There are committees for almost anything that you can think of, and for many things that you will never be able to think of. A number of faculty basically give their lives over to committee service—both at the departmental and the university level. Committees play a serious role in the governance of the institution,

---

[1]There is a variety of titles in common use for these jobs. Among them are "Vice Chairperson for Graduate Studies", "Director of Graduate Studies", "Graduate Vice Chair", and "Graduate Director". These have been used interchangeably in the present book.

from hiring to tenuring to terminating to disciplining. They are a part of the academic life.

I would be remiss not to mention that frequently the people who really keep the department going are the staff. These are the secretaries, the computer system manager, and (sometimes) the professional counselors. They are there all day every day ensuring that the department runs smoothly and that the little details that nobody ever thinks about are actually dispatched.

APPENDIX II: The Academic Ranks

In most research math departments there are five different kinds of academic positions. These are (in increasing order of importance):

- Adjunct
- Instructor
- Assistant Professor
- Associate Professor
- Professor

Let us spend some time discussing, comparing, and contrasting the different ranks.

An adjunct is a person who is hired on a temporary basis, often one course at a time, strictly to teach. This person usually does not serve on committees and has no other duties except to meet classes (and to hold office hours). The pay per course can range from $2000 to $6000 or more, depending on the school, the nature of the course, the background of the individual, and the circumstances. Usually an adjunct has no contract, no benefits, and no guarantee of future employment. Certainly there is no tenure nor any possibility of tenure. Some universities will have adjuncts on fixed-term contracts.

An Instructor (discussed also in Sections 7.2 and 7.3) or Postdoctoral Fellow is an individual, usually at a research university, who is hired to be part of the research program—often this person comes to work with a particular professor or group of professors. This is *not* a tenure-track position. It does involve teaching, although often the teaching load is reduced. It does not involve committee work or other departmental service. The position lasts two or three years but is then terminal. The position usually offers full benefits (health insurance, a retirement annuity, and so forth).

An Assistant Professor is a person on the tenure track but who is on probation. This person is not yet tenured. According to the AAUP (American Association of University Professors) rules, the Assistant Professor *must* be brought up for tenure after six years of full-time employment by the university. The procedure for tenure is described elsewhere (Sections 7.2 and 7.6) in this book. The Assistant Professor expects to become a permanent member of the faculty. These days, most research universities exercise strict standards for tenure and many Assistant Professors are denied tenure.

When an Assistant Professor is tenured, s/he is usually (although not always) promoted to the rank of Associate Professor—see Section 8.5. What is the difference between an Assistant Professor and an Associate Professor? The answer will vary from institution to institution. Usually the Associate Professor is paid a bit more. The Associate Professor will have more committee work and more responsibilities. Also, Associate Professors (along with the Professors) have a vote on future tenure cases.

Of course, an Associate Professor is a senior member of the department and is expected to take on more substantial duties. This will include directing Ph.D. theses, managing and re-thinking the curriculum, creating new courses, and participating in the hiring process.

After an Associate Professor has served in rank for about six years (this can vary considerably, depending on the institution and the candidate), s/he can be considered for promotion to Professor. The criteria are similar to the criteria for tenure and promotion to the rank of Associate Professor. The candidate must have exhibited excellence in teaching, have made real service contributions both within the department and around campus, and also have made major contributions in the research arena. Some universities will require that the candidate have begun and developed an entirely new research direction. The candidate must have made substantial service contributions, both inside the department and across campus. Outside letters are solicited and a teaching dossier is assembled. A Professor is usually paid a bit more than an Associate Professor and has more responsibilities within the department and across campus. This is the last major increment in most academic careers and the step is taken seriously. Of course Professors vote on all tenure cases and on all cases of promotion from Associate Professor to Professor.

A quite select group of full Professors is promoted to the rank of endowed Chair Professor. Usually endowed Chairs are paid an outsize salary and enjoy many other perks. The salary of an endowed Chair is paid from a special endowed fund. The endowed Chair has a reduced teaching load, travel funds, and the universal admiration and acclaim of colleagues. The endowed Chairs are the most distinguished academics on campus. They are frequently consulted by the Chancellor and other potentates. They certainly carry more than their fair share of the weight in the department.

APPENDIX III: The Academic Composition of a Mathematics Department

As indicated in Appendix II, the academic ranks of a typical American mathematics department are composed of adjuncts, Instructors/Postdocs, Assistant Professors, Associate Professors, Professors, and Chair Professors. However, the breakdown in a junior college will be different from the breakdown in a large state university, and these in turn will be different from the composition in a four-year teaching college or an elite private university. In this Appendix we shall discuss these differences. Of course no description will fit all institutions, even in one of these particular categories, but this adumbration will give you a rough idea of how departments are structured.

Junior Colleges Junior colleges do a lot of teaching, and in fairly small classes (about 25 students per class). The typical faculty teaching load is five courses per semester—as opposed to about two per semester at a research university. Junior colleges play a vital role in the community by providing education to a broad cross-section of the population. Unfortunately, these institutions are not replete with funding. As a result, a typical junior college math department has a small nucleus of permanent faculty surrounded by a very large penumbra of adjuncts. Of course there is very little scholarship or research going on at an institution like this and there are no Instructorships/Postdocs. Many junior colleges do not have the usual ranks of Assistant/Associate/Full Professors and tenure means rather a different thing at a junior college than it does at a university. It is less a form of academic protection and more a form of job security.

Since few Ph.D.'s are employed at junior colleges, we shall say no more about this topic here.

Large State Universities When I first began in this profession, about thirty years ago, there were a lot of Assistant Professors around. A typical large state math department might have had 50 or 60 tenure-track faculty, at least 10 or 15 of whom were Assistant Professors. There were also 5 to 10 Instructors. This was a healthy arrangement—speaking strictly from a scholarly point of view—because it meant a constant influx of fresh blood and new mathematical ideas.

There would also be maybe 10 or 15 Associate Professors and 20 to 30 full Professors. Well, you can imagine what happened over time. The Associate Professors (most of them) eventually became full Professors and a fair number of the Assistant Professors worked their way up through the ranks as well. Even though a certain number of the full Professors retired, the full Professor ranks got filled up.

Another interesting social phenomenon that took place in the early 1990's is that the retirement age in the United States changed from 70 years of age to infinity.[2] What does this mean? It means that now there is no mandatory retirement age for (virtually) anyone working in the United States. Of course an obvious upshot of this new law is that many people do not retire at age 70 or 75 or even at 80. It is noteworthy that, by contrast, faculty in Germany and most other European countries still must retire at about age 65.

To make a long story short, the academic world—especially at big state universities—is now well-populated with full Professors; there are far fewer Assistant Professors and just a handful of Associate Professors.[3] From the point of view of pure mathematical scholarship, this is a less healthy situation than the one we had before. Things have a greater tendency to be

---

[2]In fact the story is a bit more Byzantine than that. There was a brief transitional period during which the retirement age was raised to infinity for everyone except those in a couple of designated professions (for whom it remained 70 years of age). One of those professions was college professors. Then there came a magic moment (July 1, 1994) in which the retirement age for professors was raised from 70 to infinity. A number of faculty—including some of my colleagues—were caught on the cusp of that change and some interesting legal fisticuffs resulted.

[3]Because, obviously, the total number of tenure-track faculty is not going to go up. If anything, as recent history indicates, it will go down a bit.

moribund. It is just human nature, and the natural course of things, that people will become less mathematically active and have fewer ideas and less energy as they grow older. If a math department is composed of a preponderance of people who fit that description, and of fewer people who are young and have fire in the guts and are full of ideas, then you can imagine the result.

Fortunately, many departments have been able to hang onto a goodly number of Instructorships/Postdocs. An Instructorship is not an entitlement, so Deans like Instructorships. Put in other words, the Dean can get more instruction-for-each-dollar-spent out of an Instructor than s/he can out of a full Professor. Of course the Dean likes adjuncts even better, since they do not even get benefits (and they are paid much less per course). We must continually be on guard to keep the Dean convinced that instruction is of higher quality, and the program more robust, if the preponderance of instruction is done by tenure-track faculty. On the other hand, we must be careful not to tempt the Dean to raise our teaching loads.

Many institutions, especially the large state schools, now have attractive "Golden Parachute" programs that make it natural for tenure-track faculty to retire early—even in their early 50's. Part of the inducement is a nice financial package, and part of the inducement is that the faculty member can still keep an office and still teach from time to time. The arrangement makes sense from the institution's point of view because a retired faculty member is off the Dean's books; the retiree is paid from the State Retirement Fund.

Four-Year Teaching and Liberal Arts Colleges There are many fine scholars at the four-year teaching colleges. But the emphasis at these institutions is on teaching—teaching loads are higher, and research and scholarship play less prominent roles in tenure and promotion decisions. Institutions such as these usually do not have graduate programs—certainly they do not grant Ph.D.'s. Thus these institutions generally do not have Instructorship/Postdoc positions. They *have* been affected by the same social forces that we described in the subsection on large state universities, so the four-year colleges tend to have a preponderance of full Professors and too few Assistant Professors.

Four-year teaching colleges are usually private, so they do not have a large State Retirement Fund that they can use as a lever to create Golden Parachute programs. Thus they are perhaps harder hit by being "top heavy" than are the large state institutions. Put in harsher terms, these institutions have a harder time getting rid of tenured faculty who do not want to retire. Often the retirement package at a private institution is some combination of

Social Security (a federal program) and TIAA-CREF (a private retirement fund for teachers set up a century ago by Andrew Carnegie). The Dean can put together retirement incentives on an *ad hoc* basis, but s/he has limited resources to make any sweeping changes.

Comprehensive Universities These institutions are usually public, but they have many of the features of four-year teaching colleges. The primary mission is teaching and the school serves a broad cross-section of the population. Teaching loads are higher and research plays a less dominant role. Comprehensive universities have the usual academic ranks (Assistant Professor/Associate Professor/full Professor) and the usual tenure-and-promotion process.

Elite Private Universities Here I am talking about schools like Princeton and Harvard and Yale and MIT. These are definitely research institutions and they are heavily endowed. Like the large state schools and the four-year teaching colleges, they tend to be "tenured up", in the sense that they have a lot of full Professors and not so many Assistant Professors. They also have a retirement system—generally speaking—that is out of the Dean's control (again, usually a combination of Social Security and TIAA-CREF).

But one thing that makes the elite private math departments special is that most of them have a special endowed fund for Instructors. MIT has the Moore Instructorships, Harvard has the Peirce, Chicago has the Dicksons, and so forth. Thus, at any given moment, each of these departments may have fifteen or more Instructors on the faculty. These Instructors do a fair amount of the teaching and they also provide an influx of new ideas and energy and excitement. From my own point of view, one of the most important perks of being a member of one of the elite departments is this never-ending supply of young mathematicians.

APPENDIX IV: A Checklist for Graduate School

Here we present a checklist of things to look for when you are selecting a graduate school. You will not be able to answer all of these queries for every graduate school you are considering. If you are able to visit the school, then you will have an easier time getting the necessary information. In any event, use the list as a guide to your thoughts; let it help you sort out what you are looking for in a graduate program.

Many of the items on this checklist (the GRE, the TOEFL, financial aid, etc.) are explained or discussed in various sections of the book.

Graduate School Checklist

- [ ] Is the Math Graduate Office responsive to my needs and queries?
  - [ ] Did they send materials in a timely fashion?
  - [ ] Were materials complete and informative?
  - [ ] Did they make financial aid issues clear?
    - [ ] Does the school have a separate financial aid office?
    - [ ] Is there special financial aid for underrepresented groups?
  - [ ] Did they provide housing information?
    - [ ] Does the school have a separate housing office?
- [ ] Is the application itself clear and straightforward?
  - [ ] Is it clear how many letters of recommendation I need?

- Is it clear whether I need to take the GRE?
- Is it clear whether I need to take the TOEFL?
- Is the application deadline clear?
- Is there an application fee? How much?
- Is it clear where to send the materials?
- Is it clear when I will be notified of admission?
- Is it clear when I will be notified of financial support?
- Is it clear when I need to respond to an offer?
- What is the quality of life at this new city/school?
  - How is transportation?
    - Is the school accessible?
    - Will I need a car?
  - How is the cost of housing?
    - Are nearby apartments available and affordable?
    - Are house rentals available?
    - Is it possible, and feasible, to share living quarters?
    - Is there easy and available transportation?
  - How is the cost of food?
    - Can I eat out occasionally?
    - Are the restaurants many and varied?
  - What is the cost of insurance?
  - Is the financial aid package adequate to support a decent quality of life?
  - Is entertainment accessible and affordable?
    - Is entertainment appealing?
    - Are there major sports teams?
    - Are there golf courses, tennis courts, etc.?
    - Is there a symphony? Chamber music? A ballet?
    - Is there other good music? Rock concerts? Jazz? Ethnic music?
    - Are there good and affordable restaurants?
- Does the university provide support for underrepresented groups?
  - Is there peer counseling?
  - Are there special social events?
  - Are there professional counselors for women, African-Americans, and other groups?
  - Is the faculty diverse?
  - Is the (graduate) student body diverse?
  - Is there special financial aid?
  - Is there special housing?
- Does the university make provisions for graduate students with families?
  - Is there special financial aid?
  - Is there help in finding jobs for spouses?

APPENDIX IV

- ☐ Is there special housing?
- ☐ Is there affordable/free Day Care?
- ☐ Does the university provide for students' religious needs?
  - ☐ Is there a campus ministry?
  - ☐ Is there a Jewish student group?
  - ☐ Is there a Catholic student group?
  - ☐ Are there other student religious groups?
  - ☐ Is there a chapel on campus?
  - ☐ Are there services on campus?
  - ☐ Is there a religious social life?
- ☐ How is the camaraderie among the graduate students?
  - ☐ Are the students in competition with each other?
  - ☐ Do the graduate students study together?
  - ☐ Do the graduate students eat lunch together?
  - ☐ Do the graduate students socialize together after hours?
  - ☐ Do the graduate students support each other?
- ☐ Are the graduate students unionized?
  - ☐ If there is a union, how does it affect the quality of life?
  - ☐ What are the union dues?
  - ☐ If there is a union, is membership mandatory?
  - ☐ If there is a union, does it affect student duties?
  - ☐ If there is a union, does it affect student pay?
- ☐ Is there a graduate student government?
  - ☐ Is the student government participatory?
  - ☐ Does student government positively affect the quality of life?
  - ☐ Does the administration listen to the student leadership?
- ☐ How is the staff in this department?
  - ☐ Is there an adequate number of staff?
  - ☐ Is there one secretary dedicated to the graduate program?
  - ☐ Does the Graduate Secretary get along with the students?
  - ☐ Is the Graduate Secretary helpful to the students?
- ☐ What are the academic integrity rules at this university?
  - ☐ Are the rules clear?
  - ☐ Are the rules enforced fairly?
  - ☐ How do the rules apply to graduate students?
- ☐ What are the TA duties in this math department?
  - ☐ How many hours per week do the duties take?
  - ☐ What is the nature of the TA duties?
  - ☐ What sort of training is provided for the TAs?

- ☐ What kind of support does the department provide for TAs?
- ☐ Do I have any say in what my duties will be?
- ☐ What are the course requirements in this program?
  - ☐ What are the course requirements in the pre-qual period?
  - ☐ What are the course requirements in the post-qual period?
  - ☐ What are the overall course requirements?
  - ☐ What amount of work is required in the courses?
  - ☐ Am I allowed to audit courses?
  - ☐ How are courses graded?
- ☐ How are the qualifying exams structured at this school?
  - ☐ Is it clear how many quals I need to take?
  - ☐ Is it clear which quals I need to take?
  - ☐ Is it clear how many tries I get on the quals?
  - ☐ Is it clear how the quals are graded?
  - ☐ Is it clear how long I should take to get through the quals?
  - ☐ Are the qualifying exams, taken as a whole, of reasonable magnitude and scope?
  - ☐ Do most graduate students get through the quals in a reasonable amount of time?
  - ☐ What is the attrition rate on the quals?
- ☐ What support does the program provide immediately following the quals?
  - ☐ Is there a mechanism in place to help me find a thesis advisor?
  - ☐ Is it easy to sign up for reading courses?
  - ☐ Is it easy for me to get involved in ongoing seminars?
  - ☐ Is there a system of oral presentations or other means to get to know faculty on the research level?
  - ☐ How soon after the quals should I find a thesis advisor?
- ☐ What is the mechanism for finding a thesis advisor?
  - ☐ Are reading courses readily available?
  - ☐ Are seminars accessible?
  - ☐ Are there enough advanced and transitions courses?
  - ☐ Are there other mechanisms to get an advisor?
- ☐ What is the faculty like in this department?
  - ☐ Are faculty accessible?
  - ☐ Are faculty supportive?
  - ☐ Are faculty willing to direct theses?
  - ☐ Do faculty cooperate with each other?
  - ☐ Are faculty able to get their students jobs?
- ☐ Are seminars accessible to graduate students?
  - ☐ Are graduate students welcome in the seminars?

- ☐ Do graduate students speak in the seminars?
  - ☐ Is the seminar a natural place to find a thesis advisor?
- ☐ What is the thesis defense like?
  - ☐ Who is on my thesis committee?
  - ☐ Does the department or Graduate School help me compose the committee?
  - ☐ What is the preparation for the thesis defense?
  - ☐ What does the defense presentation consist of?
  - ☐ What will happen during the defense?
- ☐ What is the foreign language requirement?
  - ☐ Is there a written exam?
  - ☐ How extensive is the exam?
  - ☐ Is more than a reading knowledge required?
  - ☐ How many languages must I know?
  - ☐ Is previous work in a foreign language counted?
  - ☐ Which languages are allowed?
- ☐ What is the typical period of time for writing a thesis?
  - ☐ Is it made clear what constitutes "normal progress"?
  - ☐ After how many years will they threaten to cut off financial support?
  - ☐ What happens if I run out of time/money?
  - ☐ What happens if I need to change thesis advisors?
  - ☐ Will other faculty be aware of my thesis work?
  - ☐ Will fellow graduate students be aware of my thesis work?
- ☐ What technical help is there for writing the thesis?
  - ☐ Are computers available to graduate students?
  - ☐ Is TeX available to graduate students?
  - ☐ Is there instruction in using TeX?
  - ☐ Are graphics packages available?
  - ☐ Does the department provide technical typists?
- ☐ What happens after I write my thesis?
  - ☐ Will I get good advice on finding a job?
  - ☐ Will my advisor help me find a job?
  - ☐ Will I have to find my own job?
  - ☐ Will the school help me in my job search?
  - ☐ Is the program well-versed at helping people who want nonacademic employment?
    - ☐ Is there a job interview center on campus?
    - ☐ Do recruiters visit campus regularly?
    - ☐ Do the faculty have corporate and government contacts?

- [ ] Does the program keep up with people after they graduate?
  - [ ] Is there a departmental newsletter?
  - [ ] Will my advisor be there for me if I need him/her?
  - [ ] Will it be easy to keep up with my fellow students?
  - [ ] Will this program prepare me for the professional world?
  - [ ] Will I have a viable research program when I leave school?
  - [ ] How will people react when I tell them I earned my Ph.D. at this university?

Bibliography

[BEA] B. Beauzamy, Real life mathematics, *Irish Math. Soc. Bulletin* 48(2002), 43–46.

[BEC] D. Bennett and A. Crannell, eds., *Starting Our Careers*, American Mathematical Society, Providence, RI, 1999.

[BIR] L. Birnbach, *Lisa Birnbach's College Book*, Ballantine Books, New York, 1984.

[CHU] R. V. Churchill and J. W. Brown, *Complex Variables and Applications*, 6th ed., McGraw-Hill, New York, 1996.

[CON1] J. B. Conway, *On Being a Department Head, A Personal Perspective*, American Mathematical Society, Providence, RI, 1996.

[CON2] J. B. Conway, *Functions of One Complex Variable*, 2nd ed., Springer-Verlag, New York, 1978.

[DOC1] M. Do Carmo, *Differential Geometry of Curves and Surfaces*, Prentice-Hall, Englewood Cliffs, NJ, 1976.

[DOC2] M. Do Carmo, *Riemannian Geometry*, Birkhäuser, Boston, 1992.

[DUR] P. Duren, In remembrance of Allen Shields, *Math. Intelligencer* 12(1990), 11–14.

[FED] H. Federer, *Geometric Measure Theory*, Springer-Verlag, New York and Berlin, 1969.

[FOL] G. B. Folland, *Real Analysis: Modern Techniques and Their Applications*, 2nd ed., John Wiley and Sons, New York, 1999.

[GAR] T. A. Garrity, *All the Mathematics You Missed*, Cambridge University Press, Cambridge, 2002.

[**GER**] R. Geroch, *Mathematical Physics*, University of Chicago Press, Chicago and London, 1985.

[**GMF**] M. L. Goldberger, B. A. Maher, and P. E. Flattau, eds., *Doctorate Programs in the United States: Continuity and Change*, National Academy Press, Washington, D. C., 1995.

[**GRA**] G. Gratzer, *Math into LaTeX*, 3rd ed., Birkhäuser, Boston, 2000.

[**GRE**] M. J. Greenberg, *Lectures on Algebraic Topology*, Benjamin, New York, 1967.

[**GRK**] R. E. Greene and S. G. Krantz, *Function Theory of One Complex Variable*, 2nd ed., American Mathematical Society, Providence, RI, 2002.

[**HAL**] P. Hallie, *The Paradox of Cruelty*, Wesleyan University Press, Middletown, Connecticut, 1969.

[**HAG**] P. Halmos and F. Gehring, Allen Shields, *Math. Intelligencer* 12(1990), 20.

[**HER**] I. Herstein, *Abstract Algebra*, 3rd ed., John Wiley and Sons, New York, 1999.

[**JAC**] A. Jackson, New NRC rankings of graduate programs released, *Notices of the Amer. Math. Soc.* 42(1995), 1535–1540.

[**KON**] S. Kobayashi and K. Nomizu, *Foundations of Differential Geometry*, Wiley-Interscience Press, New York, 1963–1969.

[**KRA1**] S. G. Krantz, *How to Teach Mathematics*, 2nd ed., American Mathematical Society, Providence, RI, 1999.

[**KRA2**] S. G. Krantz, *A Primer of Mathematical Writing*, American Mathematical Society, Providence, RI, 1997.

[**KRA3**] S. G. Krantz, *A Handbook of Typography for the Mathematical Sciences*, CRC Press, Boca Raton, FL, 2001.

[**KRA4**] S. G. Krantz, *Real Analysis and Foundations*, CRC Press, Boca Raton, FL, 1991.

[**KRA5**] S. G. Krantz, *A Panorama of Harmonic Analysis*, Mathematical Association of America, Washington, D.C., 1999.

[**KRA6**] S. G. Krantz, *Mathematical Apocrypha*, The Mathematical Association of America, Washington, D.C., 2002.

[**LAM**] L. Lamport, *LaTeX: A Document Preparation System*, 2nd ed., Addison-Wesley, Reading, MA, 1994.

[**LAN**] S. Lang, *Algebra*, 3rd ed., Addison-Wesley, Reading, MA, 1993.

[**LOS**] L. Loomis and S. Sternberg, *Advanced Calculus*, Addison-Wesley, Reading, MA, 1968.

[MAC] S. MacLane, NRC ratings of research doctoral programs, *Notices of the Amer. Math. Soc.* 43(1996), 422–424.

[MUN1] J. Munkres, *Topology*, 2nd ed., Prentice-Hall, Englewood Cliffs, NJ, 2000.

[MUN2] J. Munkres, *Elementary Differential Topology*, Princeton University Press, Princeton, 1966.

[NRC] Reclassification of mathematics departments, *Notices of the Amer. Math. Soc.* 44(1997), 48–49.

[ONE] B. O'Neill, *Elementary Differential Geometry*, 2nd ed., Academic Press, San Diego, CA, 1997.

[RUD] W. Rudin, *Principles of Mathematical Analysis*, 3rd ed., McGraw-Hill, New York, 1976.

[SAK] S. Sawyer and S. G. Krantz, *A TEX Primer for Scientists*, CRC Press, Boca Raton, FL, 1995.

[SHA] J. Shapiro, On being Allen's student, *Math. Intelligencer* 12(1990), 15–17.

[SIS] J.-N. Silva and P. N. De Souza, *Berkeley Problems in Mathematics*, Springer-Verlag, New York, 1998.

[SPI] M. Spivak, *The Joy of TEX*, 2nd ed., American Mathematical Society, Providence, RI, 1990.

[ZDR] S. Zdravkovska, To my partner, Allen Shields, *Math. Intelligencer* 12(1990), 4–7.

Index

AAUP, 95
ability to perform graduate-level work, 9
ability to teach, 26
abstract algebra, 10
academic integrity, 86
academic integrity for graduate students, 86
academic integrity, Princeton style, 86
academic jobs, 95
academic ranks, 197
accelerated undergraduates, 11
actuarial industry, 107
actuary, 11
adjuncts, 197
administrative structure of a mathematics department, 191
administrative structure of a university, 191
admissions notification date, 23
Adobe Illustrator, 69
advanced graduate students, 5
advisor getting jobs for several students, 84
advisor, Part Zero, 33
advisor, pre-thesis, 33
advisor, preliminary, 33
Aerospace Corporation, 106
African-American students, 25
algebra at Stanford, 18
algebra, elements of, 158
ambition to be endowed Chair, 130
American Association of University Professors, 95
American Mathematical Society, 47
American Mathematical *Monthly*, 47
American Statistical Association, 47
AMS Cover Sheet, 98
AMS Employment Center, 100

AMS Employment Center, interviewers at, 101
AMS Employment Center, success in generating offers, 102
analysis at Harvard, 18
application deadline, 21
application dossier, extra materials, 23
application essay, 23
application essay, maudlin or sappy, 23
application to graduate school, confirming, 22
application to graduate school, cost of, 21
application, punctuality of, 23
applications, screening of, 21
applied math at Courant, 18
applied mathematics, 11
applying online, 20
area of study, choosing, 79
Assistant Deans, 193
Assistant Professor, life of, 111
Assistant Professors, 198
Assistant Professors as thesis advisors, 58
Assistant Professors should not write books, 113
Associate Deans, 193
Associate Professors, 198
Associate Professors as thesis advisors, 58
Association for Women in Mathematics, 47
auditing classes, 36
automotive industry, 108

baccalaureate in December, 24
basic classes at Princeton, 18
battling your way back up the ranks, 130
beasts of burden, 46
being a successful mathematician, 126

benefits associated with your job, 104
benefits of getting a problem from a senior mathematician, 60
Bergman Prize holders, 81
Berkeley system for finding an advisor, 54
Bochner, Salomon, 46
bought the barn, may as well paint it, 129
brain, banging it, 63
brain, stretching it, 63
Bulletin of the American Mathematical Society, 47

calculus, when people take, 11
career office, 109
Census Bureau, 107
central areas of mathematics, 9
Chairperson of the math department, 4
Chairperson vs. Head, 193
Chairperson, Dean selects, 193
Chairperson, Department vote for, 193
Chairperson, usually a full Professor, 193
Chairperson, who qualifies for, 193
Chairperson, who selects, 193
changing thesis advisors, 56
changing universities, 24
child prodigy as mathematician, 81
choosing a graduate school, 7
coherent body of scholarly work, 114
collaboration is like a marriage, 127
collaborators, 127
collaborators, prevalence of, 127
collaborators, value of, 127
collecting degrees, 132
college catalog, 18
college guides from Borders, 18
colloquium, 44
colloquium talk, structure of, 45
communications skills, 36
competition at the thesis level, 83
competition over jobs, 85
competition with fellow graduate students, 83
complex analysis, 10
complex analysis, elements of, 144
composition of a math department, 201
computer science, 11
concentration on teaching, 125
Concerns of Young Mathematicians, 116
concrete examples, value of, 65
confidentiality of salaries, 124
consult experienced faculty, 19
consulting, 123
Corel DRAW, 69
course on how to teach, 49
course requirement, 35, 36
courses without coursework, 36
courses, which to take, 35

criteria for graduate admissions, 17
cross-fertilization of mathematical areas, 166
cross-fertilization of teaching and research, 125
curricular preparation, 10
Curriculum Vitae, 98
CYM, 116

Daniel Wagner Associates, 106
data mining, 107
Deans, types of, 192
December baccalaureate, 24
deductive vs. inductive method, 65
denial of tenure, 130
Department Chairpersons, 193
departmental caution about English ability, 26
Departmental Office Manager, 103
departmental service as tenure criterion, 112
departmental staff, 51
departmental staff, services they can provide, 51
departmental staff, who is on?, 51
departmental tea, 44
departments tenured up, 202
depression for the mathematician, 131
Dickson Instructorships, 96
difficulty with your advisor over ideas, 63
dinner after the colloquium, 45
disenchantment with research, 125
dishonesty of mathematics professionals, 87
doing mathematics, the process of, 66
dream schools, 20
dynamical systems, 10

early graduation, 24
Educational Testing Service, 10
endowed research Instructorships, 96
energetic collaborators, 126
English ability, 26
English ability and getting a job, 27
English ability, lack of and end of mathematical career, 27
English facility and ability to teach, 27
English, you must learn, 27
entertainment, 5
essential mathematics is not written down, 65
everyone else is succeeding, 80
excellent grades, 9
excellent teaching as tenure criterion, 112

faculty diversity, 25
faculty salaries, 121
faculty who can direct theses, 58
fallback schools, 20
fellow graduate students, learning from, 43

Index

Fields Medalists, 81
filming of your teaching, 50
filtering applicants to graduate school, 12
financial aid office, 29
financial aid, sources of, 29
financial declaration for graduate school, 28
financial support in graduate school, 48
find the hottest professor around and work for him/her, 54
finding a thesis advisor before taking the quals, 54
finding a thesis advisor by asking someone cold, 56
finding a thesis advisor by reading some papers, 56
finding a thesis advisor through advanced classes, 55
finding a thesis advisor through oral exams, 55
finding a thesis advisor through reading courses, 55
finding a thesis advisor through seminars, 55
finding a thesis advisor, methods for, 55
finding a thesis problem, 59
finding a thesis problem through reading, 60
finding a thesis problem with help, 59
finding a thesis problem yourself, 59
first day of graduate school, 3
focus your work and burrow deeply, 114
focusing your attentions, 80
foreign language exam, waiving of, 47
foreign language requirement, 45
Fortran for the foreign language exam, 46
four-year colleges, emphasis on teaching, 97
four-year colleges, Instructorships at, 97
four-year colleges, jobs at, 97
four-year colleges, number of jobs at, 105
four-year colleges, tenure-track jobs at, 97
Fox sniff, 23
fraternization, rules against, 89
free ride in graduate school, 48
French, German, and Russian for the foreign language exam, 45

General Electric, 106
generosity in mathematical practice, 88
Genome Project, 107
geometry, 10
geometry/topology, elements of, 149
getting advice, 36
getting around to doing research, 126
getting fired by your thesis advisor, 85
getting people to serve on your thesis committee, 70
getting your degree at another university, 24
go to the best school you can get into, 54
going to conferences, 5

Golden Parachute programs, 203
"good old boy" network, 98
good instructors, 5
good thesis advisors, 5
grades, 17
Graduate Chairperson, 4, 193
graduate classes, effort to expend on, 37
graduate classes, level of work required, 37
graduate classes, performance expected, 37
graduate classes, work expected, 37
Graduate Committee, 4
Graduate Director, 4
graduate education, duration, 4
graduate faculty, 58
graduate program, attrition rate in, 78
graduate program, mathematical reasons for quitting, 79
graduate program, personal reasons for quitting, 78
graduate program, quitting, 78
Graduate Record Exam, 10, 12, 16
graduate school as a mistake, 131
graduate school checklist, 205
graduate school, deciding to return to, 79
graduate school, financial support, 28
graduate school, getting thrown out of, 78
graduate school, hell on wheels in, 42
graduate school, living stipend, 28
graduate school, M.D. student in, 38
graduate school, paying for, 27
graduate school, qualifications for, 22
graduate school, tuition for, 28
graphics in the thesis, 68, 69
GRE, 10, 12, 16
GRE has two parts, 17
GRE, perfect score on, 17
group photograph, 3
Gupta fiasco, 12

handbook for writing the thesis, 69
Harvard University, no Associate Professors at, 129
Harvard **Graphics**, 69
health insurance, 29
health insurance, paying for, 29
Hedrick Instructorships, 96
helping each other study for the quals, 83
Hewlett-Packard, 106
hiring someone to type your thesis, 67
holding an outside job while a student, 75
hours outside class for each hour inside, 38
how different areas of mathematics fit together, 165
how much work and how much play?, 43
how the thesis advisor selects a problem, 57
Hughes Aircraft, 106

I don't seem to know anything, 79
ideas, restructuring of, 64
impostor syndrome, 131
improving your teaching, assistance in, 49
inductive vs. deductive method, 65
industrial jobs and thesis advisors, 109
Institute for Defense Analyses, 106
Institute of Mathematical Statistics, 47
Instructor positions, 96
Instructors, 197
Instructorship, interview not required for, 101
intellectual development beyond the thesis, 94
international reputation, establishing an, 113
international students vs. American students, 43
international students, better prepared, 43
international students, comparing yourself with, 43
international students, with Master's thesis in hand, 43
intimacy with colleagues, 88
intimacy with the staff, 89
intimacy with undergraduates, 88
introduce yourself to senior faculty, 112
isolation and depression, 131

Japanese for the foreign language exam, 46
job applications, 98
job applications, number to submit, 102
job gotten with a phone call, 98
job interview center, 109
job interviews, 101
job offer, accepting, 103
job offer, getting, 103
job offer, responding to, 103
job offer, thinking about, 103
job opportunities for mathematicians, 106
job search, helping yourself, 100
jobs are mainly for young people, 105
journal, keeping a, 64

keeping your research alive, 126

lack of representation, 50
learning as a non-formalized process, 36
learning without exams, 36
learning without going to class, 36
leave, Dean's control of, 121
leave, paid, 121
leaves of absence, 79
leaving graduate school prematurely, 78
letter of acceptance of job offer, 103, 104
letters addressing mathematical attributes, 22

letters for a job application, 98
letters from mathematicians, 22
letters from outside your department, 100
letters in your tenure dossier, 115
letters of recommendation, 10, 17
letters of recommendation, asking for, 22
letters of recommendation, forged, 22
letters of recommendation, making sure they are written, 22
letters of recommendation, number of, 21–23
letters of recommendation, reminding the writers, 22
letters of recommendation, three, 21
letters of recommendation, three needed, 99
letters of recommendation, who from?, 21
letters, getting good ones for job application, 100
life of a mathematician, 133
lifelong friends from graduate school, 133
loan, getting a, 75
loan, paying off a, 75
loans from the university, 79
Los Alamos, 107

MacArthur Prize winners, 81
Major Oral, 55
majors leading to graduate school, 9
manifold theory, 10
Master's Degree as entry level, 132
Master's Degree as tarnished degree, 133
Master's Degree vs. Ph.D., 132
Master'sDegree, 132
math department composition at a four-year teaching college, 203
math department composition at a junior college, 201
math department composition at a large state university, 202
math department composition at an elite private university, 204
Mathematical Association of America, 47
mathematical growth in a new direction, 94
mathematical logic programs, 18
mathematical physics, 11
mathematical writing, 66
mathematicians can only teach, 105
mathematicians on the job market, 98
mathematics as a competitive profession, 81
mathematics of finance, 11
mathematics that you need to know, 137
`MathSciNet`, 128
medical research, 107
member dues for the union, 50
mental inertia, 65
mentoring, 36
Microsoft, 106
MikTeX, 68

mini-tenure-review, 115
Minor Oral, 55
Mitre Corporation, 106
Moore Instructorships, 96
moral turpitude, 88
most Ph.D.'s never do any additional research, 133
most Ph.D.'s stick to their thesis area, 133
multiple thesis advisors, 57

National Aeronautics and Space Administration, 108
National Security Agency, 106
Native American Students, 25
Navy, 107
nervous breakdown, having one, 44
NExT, 116
nine-month contract, 123
non tenure-track positions, 96
non-mathematical courses and job prospects, 35
non-mathematical courses, taking of, 35
nonacademic jobs, 95, 98
normal progress, 3
notes, keeping good, 64
Notices of the American Mathematical Society, 47
numerical analysis, 10

Oak Ridge, 107
old qualifying exams, studying from, 40
openness in mathematical practice, 88
oral communication, 36
organized labor among graduate students, 50
orientation procedures, 3
other sources of income, 123
outside committee members, 70
outside jobs, 75
outside jobs, regulations against, 76

paper based on thesis, 93
paper based on thesis, co-authored with advisor, 94
passion for your subject area, 53
passion for your thesis problem, 53
Peirce Instructorships, 96
Penn State Room, 128
pharmacokinetics, 107
Postdoc positions, 96
`PostScript`, 69
preparation for graduate school, 15
preparation for graduate school, necessary conditions, 15
preparation for graduate school, sufficient conditions, 16
priority disputes with thesis advisor, 63

professional societies, 47
Professors, 198
Professors as thesis advisors, 58
programmatic diversity, 25
programs that admit more students than they can graduate, 19
Project NExT, 116
promotion before tenure, 129
promotion of faculty, 128
promotion to Professor, strict criteria for, 129
prospective students, interviewing of, 26
Provost, 192
publication record, 127
publication record, importance of at four-year colleges and comprehensive universities, 128
publication record, importance of at research university, 127
publish or perish, 127
publishing while an undergraduate, 13
Putnam Exam, 14
Putnam Exam mentality, 14

qual course, work required, 37
qualifying exam graded on a curve, 83
qualifying exam graded on an absolute scale, 83
qualifying exams, 3, 9
qualifying exams at Penn State, 34
qualifying exams, Byzantine structure, 34
qualifying exams, cheating on, 86
qualifying exams, courses to prepare for, 34
qualifying exams, critical thinking skills for, 40
qualifying exams, deferring, 41
qualifying exams, failing, 40
qualifying exams, flexible structure, 34
qualifying exams, getting advice for, 34
qualifying exams, getting excused from, 35
qualifying exams, mental calisthenics for, 40
qualifying exams, passing, 39
qualifying exams, performance level required, 41
qualifying exams, preparation for, 34, 35
qualifying exams, quality of work on, 41
qualifying exams, second try on, 41
qualifying exams, standardized set of, 33
qualifying exams, strict grading of, 42
qualifying exams, studying for, 39
qualifying exams, the nature of, 39
qualifying exams, too hard, 34
qualifying exams, too long, 34
qualifying exams, typical questions on, 39
qualifying exams, what you must know for, 40
qualifying exams, when to complete, 33

qualifying exams, when to take, 33, 40
qualifying exams, where you stand, 35
quality of job you can expect, 104

RA support for graduate students, 51
RAND Corporation, 106
reading of the thesis, 70
real analysis, 10
real analysis, elements of, 138
realistic schools, 20
recitation sections vs. lectures, 48
recitation sections, teaching of, 48
records, keeping good, 65
reform teaching, 99
research as tenure criterion, 113
Research Assistant support for graduate students, 51
research faculty, 58
research universities, number of jobs at, 105
respect for august faculty, 59
reviewing for publishers, 123
rules for writing the thesis, 69

sabbatical, applying for, 120
sabbaticals, 119
sabbaticals at the University of California, 120
sabbaticals only for tenure-track faculty, 120
sabbaticals, competition for, 120
safety schools, 20
salary of a Professor, 122
salary of a Research Instructor, 122
salary of an Adjunct Instructor, 121
salary of an Assistant Professor, 122
salary of an Associate Professor, 122
sanctity of your ideas, 63
seminars, 126
senior faculty mentor, 115
sex with colleagues, 88
sexual harassment, 88
Shields method for directing a thesis, 62
Silver Spoon College, 11
smoking gun for promotion to Professor, 129
Social Security Administration, 108
Society for Industrial and Applied Mathematics, 47
solicitation of tenure letters, 115
solving thesis problem, nonlinear path to, 64
solving your original thesis problem, 61
specializing so that you can finish, 79
Sputnik era, 4, 118
St. Olaf College, 12
staff keeps the Department going, 195
statistician, 11
stealing of your ideas, 63
stochastic processes, 11
student voice, 50

summer research grants, 123
survival after denial of tenure, 130
Sylvania, 106
synthesis of mathematical areas, 137, 166
systems analysis, 107

TA duties, 5, 48
TA training, 49
taking courses in different fields, 167
talk to people, 18
talking about your problem, the value of, 65
talking to other faculty besides the thesis advisor, 56
talking to people, 36
talking to your thesis advisor, 85
Teaching Assistant, 3
Teaching Assistant duties, 48
Teaching Center, 49
teaching dossier, 95, 99
teaching duties, 26
teaching duties, taking them seriously, 48
teaching experience and looking for a job, 26
teaching load, 117
teaching loads at elite private universities, 119
teaching loads at four-year teaching colleges, 119
teaching loads at junior colleges, 118
teaching loads at large state universities, 118
teaching seminar, 49
Teaching Statement, 98
teaching statistics, 99
teaching with computers, 99
tenure appeals, 95
tenure document, 128
tenure in Florida, 128
tenure in Georgia, 129
tenure is for life, 96
tenure process, 95
tenure track, 96
tenure, Board of Trustees adjudication of, 95
tenure, Chancellor's adjudication of, 95
tenure, Dean's adjudication of, 95
tenure, departmental adjudication of, 95
tenure, meaning of, 95
tenure, Provost's adjudication of, 95
tenure-track job, interview required for, 101
tenure-track positions, fresh Ph.D.'s not qualified for, 96
tenure-track positions, qualifications for, 96
TeX, flavors of, 67
TeX, learning of, 68
TeX, obtaining the software, 68
TeX, use of, 67
Texas Instruments, 106
thesis advisor, 4
thesis advisor and thesis problem, 53

Index

thesis advisor at another university, 24
thesis advisor steers you to a problem, 57
thesis advisor that does not work out, 56
thesis advisor with several graduate students, 84
thesis advisor, choosing, 53
thesis advisor, collaboration with, 94
thesis advisor, comfortable relationship with, 62
thesis advisor, getting advice from, 62
thesis advisor, getting consolation from, 63
thesis advisor, getting fired by, 85
thesis advisor, getting help from, 62
thesis advisor, how to get in trouble with, 85
thesis advisor, no appointments with, 61
thesis advisor, personal relationship with, 62
thesis advisor, prepare to talk to, 62
thesis advisor, regular appointments with, 61
thesis advisor, what to call, 59
thesis advisor, your relationship with, 61
thesis advisors who are "nice guys", 54
thesis committee, 70
thesis committee, composition of, 70
thesis defense, 70
thesis defense proceedings, 71
thesis defense, failing, 72
thesis defense, Swedish method, 72
thesis problem, adjusting conclusions, 77
thesis problem, adjusting hypotheses, 77
thesis problem, adjusting so you can solve, 76
thesis problem, blockage on, 64
thesis problem, cannot solve, 76
thesis problem, changing, 61
thesis problem, finding, 59
thesis problem, frustration with, 64
thesis problem, how to work on, 63
thesis problem, how you know when you have solved it, 78
thesis problem, not in a book, 63
thesis problem, not solving your first, 61
thesis problem, solving, 60
thesis problem, solving by increments, 64
thesis problem, stuck on, 76
thesis problem, working by experimentation, 64
thesis to be submitted in hard-copy, 70
thesis, all arguments complete, 66
thesis, cheating on, 87
thesis, dishonesty with, 87
thesis, first outline for, 66
thesis, form in which to publish, 93
thesis, gaps in the argument, 67
thesis, how much is enough for?, 77

thesis, how to publish, 93
thesis, how to tell when it is done, 77
thesis, protecting the ideas in, 87
thesis, publishing of, 93
thesis, second outline for, 66
thesis, the writing process, 67
thesis, where to publish, 93
thesis, who on the committee will read, 70
thesis, writing of in $\TeX$, 67
thesis, writing up of, 66
things you can do to find a job, 100
three letters required, 98
time in rank of Associate Professor, 130
TOEFL Exam, 26
tooling up, 80
top-rated graduate schools, 21
topology, 10
training to be a TA, 49
tutoring for extra money, 76
twelve-month contract, 123

UCSC, 16
undergraduate education, duration, 4
undergraduate education, gaps in, 40
undergraduate research, 13, 14
undergraduate research and eligibility for graduate school, 13
undergraduate, one of the best, 42
underrepresented groups, 25
underrepresented groups, special fellowships for, 25
underrepresented groups, special housing for, 25
underrepresented groups, special mentors for, 25
unionization of graduate students, 50
University of California at Santa Cruz, 16
University of California step system, 124
upper division math courses, 9

Vice Chairperson for Graduate Studies, 194
Vice Chairperson for Undergraduate Studies, 194
Vice Chairpersons, 194
visit for candidates, 24
vita, 98

what to call your thesis advisor, 59
when show up for first job, 112
women students, 25
women, isolated in graduate school, 25
work habits, good, 64
work, throwing away 90% of, 64
work, you have to be willing to do it, 86
working all the time, 43
working with more than one thesis advisor, 57

writing books, 113, 123
writing up the thesis, 66